LIFE'S GRANDEUR

LIFE'S GRANDEUR

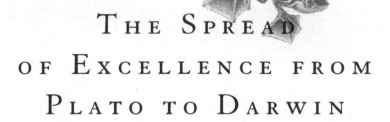

THE SPREAD
OF EXCELLENCE FROM
PLATO TO DARWIN

STEPHEN JAY GOULD

JONATHAN CAPE
LONDON

First published 1996

1 3 5 7 9 10 8 6 4 2

© Stephen Jay Gould 1996

Stephen Jay Gould has asserted his right
under the Copyright, Designs and Patents Act 1988
to be identified as the author of this work

First published in the United States in 1996 by Harmony Books as *Full House*

First published in the United Kingdom in 1996 by Jonathan Cape,
Random House, 20 Vauxhall Bridge Road, London SW1V 2SA

Random House Australia (Pty) Limited
20 Alfred Street, Milsons Point, Sydney,
New South Wales 2061, Australia

Random House New Zealand Limited
18 Poland Road, Glenfield,
Auckland 10, New Zealand

Random House South Africa (Pty) Limited
Box 2263, Rosebank 2121, South Africa

Random House UK Limited Reg. No. 954009

A CIP catalogue record for this book is available from the British Library

Papers used by Random House UK Limited are natural,
recyclable products made from wood grown in sustainable forests.
The manufacturing processes conform to the environmental
regulations of the country of origin

ISBN 0–224–04132–0

Printed and bound in Great Britain
by Mackays of Chatham PLC

For Rhonda,

who is the embodiment of excellence

• • •

Das Ewig – Weibliche

zieht uns hinan

•Contents•

Part One
• • •

How Shall We Read and Spot a Trend?

Part Two
• • •

Death and Horses:
Two Cases for the Primacy of Variation

PART THREE
• • •

The Model Batter:
Extinction of 0.400 Hitting
and the Improvement of Baseball

PART FOUR
• • •

The Modal Bacter:
Why Progress Does Not Rule
the History of Life

LIFE'S GRANDEUR

• A Baseball Primer for British Readers •

In our increasingly fragmented and parochial world, few phenomena other than global wars, pandemic diseases, and the Olympic Games bring us all together for common purposes. This book, written from an American parish, uses baseball—the quintessential shibboleth of my culture—as one of two central examples to carry the major message. This strategy may be terrific for Yanks, but what a turnoff for Brits! (I'd be truly pissed off if Stephen Hawking based his next book on grasping an analogy between the structure of the universe and hitting for six, bowling a maiden over, or being out leg before wicket.) Consequently, I hasten to provide this synopsis of an arcane American religion. (Baseball, of course, is so deep, so rich, and so subtle that this meagre effort can only be as absurd as a ten-page, easy-reading, comic book version of the *Summa Theologica*. But, as they say, once more unto the breach . . . .)

America is too young for mythic heroes. We have no distant King Arthurs, and must therefore invest our legends in real people who slay British tyrants (George Washington), free slaves (Abraham Lincoln), or emerge from an orphanage to hit sixty home runs in a single season (Babe Ruth). Baseball, a genuine sport that must also serve double duty (in this context) as a primal mythic institution, evolved in nineteenth-century America from various English stick-and-ball games (Jane Austen mentions something called "base ball" in a late eighteenth-century novel). One of our two modern professional leagues began in 1876, the other in 1901. Baseball gains its mythic and ecumenical status (within American culture) by virtue of its age and its original constitution as a pastime for all people, centred in rural and industrial urban life (whereas American football began in universities at a time when few people pursued tertiary education, while basketball arose a good deal later and remained, until recently, a more restricted and largely indoor sport).

Two other peculiarities of baseball's history and structure abet the mythology and make such writings as Part Three of this book possible. First, baseball has experienced no major change of rules since 1893. Thus, events of a distant past are truly comprehensible and comparable with modern accomplishments. Second, although baseball is a team sport, each major action is a contest between individuals (pitcher against batter, runner against fielder, etc.). Consequently, statistics for personal performance have clear meaning and comparability (whereas passes attempted in football, or points scored in basketball, depend so crucially upon a team's overall strategy that we cannot meaningfully compare individual performances across teams and times). The lore of baseball is therefore awash in statistics. Any serious fan can tell you how many homers the Babe hit in 1927, how many games Cy Young won during his career, and how many ribbies that little stump of a man, Hack Wilson at five feet six, got in 1930. Such arcana immediately pose an insurmountable problem in translation. I can tell you the basic rules of the game in a few pages, but I can't transmit the lore—for this kind of "feel" requires a lifetime of involvement. Thus, you may end up understanding the points I raise in Part Three, but still be absolutely and utterly mystified as to why anyone would ever give a sliver of damn. Here I can only preach tolerance for national idiosyncracy. I am equally befuddled as to why a chorus of ecstacy should accompany the report that W. G. Grace, in his last match at age sixty-six, scored sixty-nine not out for Eltham. But I do accept that this is important—and I would as soon disparage this figure as I would tell Jesus or John the Baptist to get a shave and change clothes.

A baseball field (see accompanying diagram) has a diamond-shape infield with four bases at the corners, and a wedge-shaped outfield beyond. Balls hit into the wedge and beyond are "fair" and in play; balls hit to the side are "foul" and not in play. The four bases, proceeding counterclockwise as players must run, are called home plate, first base, second base, and third base. (This terminology will help you understand American slang. When a young stud says that "he couldn't get to first base" with his date, you will know to applaud her fortitude and appreciate his frustration.) The batter stands at the home plate, and the pitcher throws (he does not bowl!) from the middle of the infield past the batter (or so he hopes) and to the catcher, crouching behind home plate. The other seven fielders (for a baseball team has nine players) arrange themselves as follows: four infielders (first, second and third basemen, and a shortstop who plays between second and third base because most batters are righthanded and hit more balls this way than between first and second bases); and three outfielders (in left, center and right field—with left and right defined by the batter looking outward from home plate).

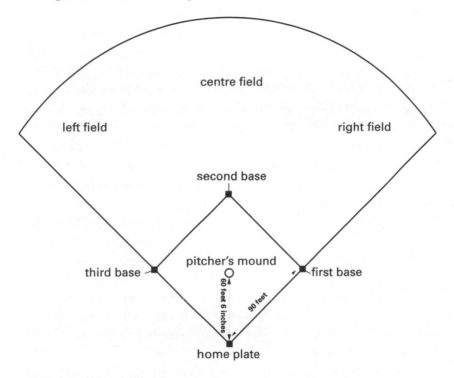

As in most games, the object of play is to score more points (called runs in baseball) than the other team in the allotted duration. Baseball, unlike most team sports, does not define duration of play by clock time at all (although the average game lasts about three hours). Each side comes to bat (alternating with the other side) nine times (called innings, but not otherwise particularly comparable to the fewer and longer items of the same name in cricket). Your team's part of the innings continues until three of your men have been put "out" (to be defined in a moment). The game ends after each side has completed its nine innings (at three outs per inning for a total of twenty-seven outs). The team with more runs wins the game.

The actual procedure is pretty primal. The batter tries to hit safely and run around the bases. The pitcher and his fielders try to put the batter out. A batter may be put out in the following major ways (not including lbw!): if the ball he hits is caught before it touches the ground (called a fly-out); if a ball that hits the ground is snagged by a fielder and thrown to the first baseman before the batter can run to first base; if the pitcher manages to throw three good pitches (called "strikes") past the batter (called a strike-out). "Three strikes and you're out" is therefore a mantra of American culture—something you Brits really need to know if you hope to understand all manner of things American, including Gary Cooper's famous movie line when, playing baseball hero Lou Gehrig, he learns of his imminent death from ALS, now called Lou Gehrig's Disease: "Is this strike three, doc?" In our current climate of conservative backlash, several American states have instituted so-called "strike-three laws" mandating life sentences without possibility of parole for third-time offenders. And how these reprobates must curse the historical contingency that, so long ago and for other purposes, specified three strikes rather than five or six for an out!

If a batter hits safely (usually a ground ball that gets past the infielders and rolls into the outfield, or a fly ball that falls between outfielders), he runs as far as he can—reaching either first base (called a single), second base (a double) or third base (a triple). (The batter stops running when a fielder grabs the ball and throws it to another fielder covering the base just past the batter's last advance—for if the ball reaches a base before the batter does, and the fielder can "tag" the batter with the ball, then the batter is out. Thus, if the batter reaches second base and judges that he cannot get to third base before the fielded ball, he will stay put.) A run does not score until the batter manages to run around all the bases and reach home plate. Thus, if the batter chooses to stop at second base, he remains there until a subsequent batter hits the ball and permits him to advance. (All manner of rules and customs govern the wisdom and possibility of a baserunner's advancement when a subsequent batter hits the ball—but this we must leave to more learned treatises.)

In the most honored feat of all—another American icon demanding reverential obeisance—a batter may achieve the equivalent of your hitting for six by hitting a fair ball beyond the outfield on the fly and into the spectator's gallery beyond, where a lucky spectator (called a fan), in another time-honored ritual, gets to keep the ball. Such a shot is called a "home run," or a "homer," or a "dinger," or a "round tripper," or a hundred other things (some unprintable and uttered by the pitcher who served up the ball). The batter who hits a home run scores one run for himself, and an additional run for any teammate then occupying another base by virtue of a previous hit—up to a maximum of four runs if the bases are "loaded" (that is, one of your guys on each base), for a so-called "grand slammer". Unlike cricket, where you don't even get to move or brag after hitting for six, any batter who hits a home run then follows the grand ritual of trotting (usually very slowly, for maximal effect) around all the bases in order (sometimes throwing a bird to the pitcher and receiving the finger in return), until he crosses home plate, where all his teammates converge for handshakes and high fives.

I don't want to drown you in details, but I must add (for completion) another way or two for becoming a baserunner without hitting the ball. Three strikes, you're out; but four balls, you're on. If the pitcher throws four errant pitches (outside a small area around home plate and between the batter's belt and knees, called the "strike zone"), then the batter moves to first base with a "walk" or a "base on balls". (Yes, I know, the sphere thrown by the pitcher is a ball. But only errant pitches are called "balls". Accurate pitches are called "strikes". If you found this confusing, you will have to complain to higher powers than this poor author.) In another crucial motto (with purely practical rather than moral meaning), "a walk is as good as a hit"—for a baserunner on first base is a baserunner on first base no matter how he got there: that is, he will score the same run (if subsequent batters advance him all the way around) whether he got to first base by walking or hitting. A batter also goes to first base if he is hit by an errant pitch—baseball's only real defense against perpetual mayhem. (We are not civilized enough to call a batter out lbw if he gets in the way of a pitch.)

Well folks, that's pretty much it. But my library contains eleven large shelves full of baseball books—so there's a lot of history and subtlety that I've left out. You will be on the path to understanding when you grasp the major structural difference between baseball and cricket: in baseball, you must run (and be either safe on a base or out) whenever you hit the ball into fair territory. That is, each safely batted ball must result in either a hit or an out. This custom makes baseball go ever so much faster than cricket (and resolves, for you diehard cricketers, the apparent absurdity that a team could play nine full innings of anything in just a few hours). Baseball seems slow and boring to many hyped-up Americans in the modern age of sound-bite culture. But baseball moves with the wind compared to a game that gives you the option of a null result—no running and no possible run or out—when you hit the ball, thereby mandating a strategy of time-killing by dribbling deflection in certain circumstances. Hey, don't get me wrong. I love cricket. I also love *Parsifal*.

This entire disquisition finally leads me to the point of all this—an explanation of three key statistics (one for batting, one for fielding, and one for pitching) that measures performance in baseball's three major activities, and that form the basis for my argument in Part Three:

BATTING AVERAGE: A player's batting average is simply his ratio of hits to total times at bat (walks don't count as an official time at bat, for a hitter shouldn't get credit for a pitcher's malfeasance, but he hasn't failed either). Thus, a batting average of 0.300 (considered excellent by the way, and reached by fewer than 10 percent of players each year) means that, on average, a batter has gotten three hits and made seven outs for each ten times at bat. (In another baseball maxim, we are fond of saying that baseball is the only sport where the best players succeed in fewer than one-third of their attempts.) A batting average of 0.400 indicates four safe hits in every ten times at bat. No one has hit higher than 0.400 in major league baseball since 1941—although seven players reached this level between 1900 and 1930. Part Three uses the key argument of this book to prove, contrary to all voluminous prior discussion of this historical pattern, that (paradoxically perhaps) the disappearance of 0.400 hitting actually measures the general improvement of play in baseball. Read on.

FIELDING AVERAGE: If a batter reaches a base safely because a fielder drops a fly ball that he should have caught, or bobbles a ground ball that he should have snared, or throws a ball errantly to another fielder, then the guilty fielder has committed an "error". The fielding average is simply the percentage of balls handled properly. Since fielders are damned good these days, fielding averages tend to measure near the maximum of 1.000—or all balls handled properly. A fielding average of 0.990—often achieved by the way—really does mean that a fielder has handled 99 of 100 balls accurately.

EARNED RUN AVERAGE (or ERA): This fundamental measure of pitching prowess is simply the average number of "earned" runs scored against a pitcher in a full nine innings. (Thus, an ERA of 2.0—damned good and rarely achieved—means that a pitcher has given up an average of 2 runs to the other side in each full game.) "Earned" runs are those that can be charged to the pitcher's malfeasance. It would not be fair to blame the pitcher, after all, if a fielder dropped a ball that should have been the third out and the opposing team than went on to score runs—for the pitcher's proper work should have ended the inning with no further runs. (As a general measure of effectiveness in pitching, we prefer the ERA to the simple total of games won, to the ratio of games won to games lost—for the exact same pitching performance will win fewer games for a lousy team than for a terrific team that backs you up with good hitting and a pile of runs.)

At present, professional baseball maintains two major leagues, each with three divisions. Each team plays a season (April to early October) of 162 games. Two rounds of playoffs follow to determine the champion of each league. The two champions then meet for a best-of-seven set of games (ending when one team scores four victories) called, in our greatest parochialism of all, the World Series. Yes, not *a* World Series, but *The* World Series. And yes again, grown (and reasonably intelligent) people do take this stuff seriously. I have just spent a lovely afternoon at a type-writer telling you why—and I only scratched the surface. Any religion looks nutty to outsiders, but there must be something to it.

PS: Although I am confident that poker has crossed the Atlantic far more efficiently than baseball, I still hesitate to use the American title of this book, *Full House* (a good poker hand expressing both high value and use of all items—that is, the full range of variation). Consequently, I turn instead to my all-time favorite Englishman, Charles Darwin, and adapt his equally appropriate final statement from the *Origin of the Species*— "there is grandeur in this view of life…"—as a title for the British edition. I mention this not as an agent for Las Vegas, or as a general shill for American pastimes, but only so that British readers will not be mystified by numerous repetitions of the phrase "full house" in the text of this book. I use this poker metaphor to emphasize my central theme that we can only understand trends properly if we map expansions and contradictions in variation among all items in systems, and cease to focus on the march of mean or extreme values through time.

• A Modest Proposal •

In an old literary theme, from Jesus' parable of the prodigal son to Tennessee Williams's *Cat on a Hot Tin Roof,* our most beloved child is often the most problematic and misunderstood among our offspring. I worry for *Full House,* my adored and wayward boy. I have nurtured this short book for fifteen years through three distinctly different roots (and routes): (1) an insight about the nature of evolutionary trends that popped into my head one day, revised my personal thinking about the history of life, and emerged in technical form as a presidential address for the Paleontological Society in 1988; (2) a statistical eureka that brought me much hope and comfort during a life-threatening illness (see chapter 4); and (3) an explanation that, once conceptualized, struck me as self-evident and necessarily correct, but also diametrically opposed to all traditional accounts, for a major puzzle of American popular culture—the disappearance of 0.400 hitting in baseball.

All three roots arose from a common insight in the form most personally exciting to intellectuals—the eureka or *a-ha!* moment that inverts an old way of seeing and renders both clear and coordinated something that had been muddy, inchoate, or unformulated before. (I speak of a deeply personal experience, not a claim full of hubris about absolutes. Such eurekas only remove scales from one's own eyes and break idiosyncratic impediments. The rest of the world may always have known what you just discovered. But then, some eurekas are more generally novel.) My insight made me view trends in an entirely different way: as changes in variation within complete systems, rather than as "a thing moving either up or down" (hence the subtitle of this book, The *Spread* of Excellence).

With insight came fear—and for two reasons. First, the theme may seem small and offbeat at first. Why should a different explanation of trends become a subject of general interest? Moreover, and second, the key reformulation (thinking of whole systems expanding or contracting, rather than entities on the move) is fundamentally statistical and must be presented in graphical terms. I did not fear for incomprehensibility. The key idea is as simple as could be (a conceptual inversion, not an arcane mathematical expression), and I knew that I could present the argument entirely in pictorial (not algebraic) terms. But I also knew that I would have to lay out the argument carefully, first making the general point and then developing some simple and preliminary examples before taking on the two main subjects: 0.400 hitting and a resolution of the problem of progress in the history of life.

But would people read the book? Would readers persist through the necessary preliminaries to reach the key reformulations? Would they maintain interest through a graphical development, given our cultural disinclination toward anything that smacks of mathematical style? Yet, I remain convinced that this book presents a novel argument of broad applicability—and that persistent readers may emerge with satisfaction, and in agreement with the father as he pardoned his prodigal son (and justified mercy to his other, persistently obedient child): "it was meet that we should make merry and be glad."

So let me make a deal with you. As a man who has spent many enlightening, if unenriching, hours playing poker (hence the book's title), I want to propose a bet. Persist through to the end, and I wager that you will be rewarded (perhaps even with a royal flush to beat my full house). In return, I have made the book short (remarkably so compared with my other effusions), hopefully clear and entertaining (if methodical in building up to the two main examples), and imbued with a promise that two truly puzzling, important, and apparently unrelated phenomena can be explained by the conceptual apparatus here developed.

The rewards of persistence should be twofold. First, I think that my approach of studying variation in complete systems does provide genuine resolution for two widely discussed issues that can only remain confusing and incoherent when studied in the traditional, persistently Platonic mode of representing full systems by a single essence or exemplar—and then studying how this entity moves through time. I find both resolutions particularly satisfying because they are not so radical that they lie outside easy conceivability. Rather, both solutions make eminent good sense and resolve true paradoxes of the conventional view, once you imbibe the revised perspective based on variation. How can we believe, as the traditional approach requires, that 0.400 hitting has disappeared because batters have gotten worse, when record performances have improved in almost any athletic activity? My approach shows that the disappearance of 0.400 hitting actually records the increasing excellence of play in baseball—and this makes satisfying sense (but cannot be coherently grasped at all under traditional modes of thought about the problem).

Similarly, although I can marshal an impressive array of arguments, both theoretical (the nature of the Darwinian mechanism) and factual (the overwhelming predominance of bacteria among living creatures), for denying that progress characterizes the history of life as a whole, or even represents an orienting force in evolution at all—still, and if only for legitimate parochial reasons, we rightly embrace the idea that humans are uniquely complex, and we properly insist that this fact requires some acknowledgment of a trend. But the explanatory apparatus of *Full House* permits us to retain this commonsensical view about human status, while understanding that progress truly does not pervade or even meaningfully mark the history of life.

Second—and I don't quite know how to say this without sounding more immodest than I truly intend to be—this book does have broader ambitions, for the central argument of *Full House* does make a claim about the nature of reality. I say nothing that has not been stated before by other folks in other ways, but I do try to explicate a broad range of cases not usually gathered together, and I am making my plea by gentle example, rather than by tendentious frontal assault in the empyrean realm of philosophical abstraction (the usual way to attack *the* nature of reality, and to guarantee limited attention for want of anchoring). I am asking my readers finally and truly to cash out the deepest meaning of the Darwinian revolution and to view natural reality as composed of varying individuals in populations—that is, to understand variation itself as irreducible, as "real" in the sense of "what the world is made of." To do this, we must abandon a habit of thought as old as Plato and recognize the central fallacy in our tendency to depict populations either as average values (usually conceived as "typical" and therefore representing the abstract essence or type of the system) or as extreme examples (singled out for special worthiness, like 0.400 hitting or human complexity). The subtitle of this book—The Spread of Excellence from Plato to Darwin—epitomizes the two approaches, and the importance of owning Darwin's solution.

Full House is a companion volume of sorts to my earlier book *Wonderful Life* (1989). Together, they present an integrated and unconventional view of life's history and meaning—one that forces us to reconceptualize our notion of human status within this history. *Wonderful Life* asserts the unpredictability and contingency of any particular event in evolution—and emphasizes that the origin of *Homo sapiens* must be viewed as such an unrepeatable particular, not an expected consequence. *Full House* presents the general argument for denying that progress defines the history of life or even exists as a general trend at all. Within such a view of life-as-a-whole, humans can occupy no preferred status as a pinnacle or culmination. Life has always been dominated by its bacterial mode.

Both volumes present their basic arguments through particular examples (of an arresting sort), rather than by tendentious generalities—the full range of the Cambrian explosion as revealed in the fauna of the Burgess Shale in *Wonderful Life*; the disappearance of 0.400 hitting in baseball, and the constant bacterial mode of life's bell curve in *Full House.* These cases suggest that we trade the traditional source of human solace in separation for a more interesting view of life in union with other creatures as one contingent element of a much larger history. We must give up a conventional notion of human dominion, but we learn to cherish particulars, of which we are but one *(Wonderful Life),* and to revel in complete ranges, to which we contribute one precious point *(Full House)*—a good swap, I would argue, of stale (and false) comfort for broader understanding. It is, indeed, a wonderful life within the full house of our planet's history of organic diversity.

So you have my modest proposal. Please read this book. Then let's talk, and have a whale of an argument about all manner of deepest things—and of cabbages, and kings.

Part One

• • •

How Shall
We Read and Spot
a Trend?

· 1 ·

Huxley's Chessboard

We reveal ourselves in the metaphors we choose for depicting the cosmos in miniature. Shakespeare, unsurprisingly, saw the world as "a stage, and all the men and women merely players." Francis Bacon, in bitter old age, referred to external reality as a bubble. We can make the world really small for various purposes, ranging from religious awe before the even grander realm of God ("but a small parenthesis in eternity" according to Sir Thomas Browne in the mid-seventeenth century), to simple zest for life (as stated so memorably in a conversation between the paragons for such a position, Pistol and Falstaff: "the world's mine oyster, which I with sword will open").

We should therefore not be surprised that Thomas Henry Huxley, the arch rationalist and master of combat, should have chosen a chessboard for his image of natural reality:

> The chess board is the world, the pieces are the phenom-
> ena of the universe, the rules of the game are what we call
> the laws of Nature. The player on the other side is hid-
> den from us. We know that his play is always fair, just,
> and patient. But also we know, to our cost, that he never
> overlooks a mistake, or makes the smallest allowance for
> ignorance. (From *A Liberal Education,* 1868.)

This image of nature as a tough but fair adversary, beatable by the two great weapons of observation and logic, underlies Huxley's most famous pronouncement that "science is simply common sense at its best; that is, rigidly accurate in observation and merciless to fallacy in logic." (From his great popular work *The Crayfish,* 1880.)

Huxley's metaphor fails—and our task in revealing nature becomes cor-respondingly harder—because we cannot depict the enterprise of science as Us against Them. The adversary at the other side of the board is some com-plex combination of nature's genuine intractability and our hidebound so-cial and mental habits. We are, in large part, playing against ourselves. Nature is objective, and nature is knowable, but we can only view her through a glass darkly—and many clouds upon our vision are of our own making: social and cultural biases, psychological preferences, and mental lim-itations (in universal modes of thought, not just individualized stupidity).

The human contribution to this equation of difficulty becomes ever greater as the subject under investigation comes closer to the heart of our practical and philosophical concerns. We may be able to apply maximal objectivity to taxonomic decisions about species of pogonophorans in the Atlantic Ocean, but we stumble in considering the taxonomy of fossil human species or, even worse, the racial classification of *Homo sapiens.*

Thus, when we tackle the greatest of all evolutionary questions about human existence—how, when, and why did we emerge on the tree of life; and were we meant to arise, or are we only lucky to be here—our preju-dices often overwhelm our limited information. Some of these biased de-scriptions are so venerable, so reflexive, so much a part of our second nature, that we never stop to recognize their status as social decisions with radical alternatives—and we view them instead as given and obvious truths.

My favorite example of unrecognized bias in depicting the history of life resides quite literally in the pictures we paint. The first adequate reconstructions of fossil vertebrates date only from Cuvier's time, in the early nineteenth century. Thus the iconographic tradition of drawing successive scenes to illustrate the pageant of life through time is not even two centuries old. We all know these series of paintings—from a first scene of trilobites in the Cambrian sea, through lots of dinosaurs in the middle, to a last picture of Cro-Magnon ancestors busy decorating a cave in France. We have viewed these sequences on the walls of natural history museums, and in coffee-table books about the history of life. Now what could be wrong, or even strongly biased, about such a series? Trilobites did dominate the first faunas of multicellular organisms; humans did arise only yesterday; and dinosaurs did flourish in between.

Consider three pairs of scenes spanning a century of this genre, and including the three most famous practitioners of all time. Each shows a Paleozoic and a Mesozoic marine scene. In each, the Paleozoic tableau features invertebrates, while the Mesozoic scene shows only marine reptiles that have descended from terrestrial forms. The first pair comes from a work that established the genre in the early 1860s—Louis Figuier's *La terre avant le déluge* (*The World Before the Deluge;* see Rudwick's fascinating book, *Scenes from Deep Time,* for a survey of this genre's foundation in the nineteenth century). The second is the canonical American version, painted by Charles R. Knight, greatest artist of prehistoric life, for an article in *National Geographic Magazine* (February 1942), and titled *Parade of Life Through the Ages.* The last pair represents the equally canonical European work of the Czech artist Z. Burian in his 1956 work written with paleontologist J. Augusta and entitled *Prehistoric Animals.*

So why am I complaining? No vertebrates yet lived in the early Paleozoic, and marine reptiles did return to the sea during dinosaur days in the Mesozoic. The paintings are "right" in this narrow sense. But nothing can be more misleading than formally correct but limited information drastically yanked out of context. (Remember the old story about the captain who disliked his first mate and recorded in the ship's log, after a unique episode, "First mate was drunk today." The mate begged the captain to remove the passage, stating correctly that this had never happened before and that his

9

employment would be jeopardized. The captain refused. The mate kept the next day's log, and he recorded, "Captain was sober today.")

As for this nautical tale, so for the history of life. What can be more misleading than the representation of something small as everything typical? All prominent series of paintings in this genre of prehistoric art—there are no exceptions, hence the example's power—claim to be portraying the nub or essence of life's history through time. They all begin with a scene or two of Paleozoic invertebrates. We note our first bias even here, for the prevertebrate seas span nearly half the history of multicellular animal life, yet never commandeer more than 10 percent of the pictures. As soon as fishes begin to prosper in the Devonian period, underwater scenes switch to these first vertebrates—and we never see another invertebrate again for all the rest of the pageant (unless a bit-playing ammonite squeezes into the periphery of a Mesozoic scene). Even the fishes get short shrift (literally), for not a single one ever appears again (except as fleeing prey for an ichthyosaur or a mosasaur) after the emergence of terrestrial vertebrate life toward the end of the Paleozoic era.

Now, how many people have ever stopped to consider the exceedingly curious and unrepresentative nature of such limited pageantry? Invertebrates didn't die or stop evolving after fishes appeared; much of their most important history unrolls in contemporaneous partnership with marine vertebrates. (For example, the most fascinating and portentous episodes of life's history—the five largest mass extinctions—are all best recorded by changes in invertebrate faunas.) Similarly, fishes didn't die out or stop evolving just because one lineage of peripheral brethren managed to colonize the land. To this day, more than half of all vertebrates are fishes (more than 20,000 living species). Isn't it absurd to eliminate a vertebrate

FIGURE 1

Three paired views of artistic representations of the history of life to show the unchanging biases that pervade this genre. The three pairs come from the work of Figuier in the 1860s, Knight in the 1940s, and Augusta and Burian in 1956. The first member of each pair shows invertebrates from the early history of multicellular life. The second figure in each pair shows a marine scene from the Mesozoic Era (time of domination of dinosaurs on land). No fish or invertebrates are shown in the Mesozoic scene, but only reptiles that have returned to the marine environment.

majority from all further pictorial representation just because one small lineage changed its abode to land?

The story of terrestrial vertebrates is just as egregiously biased. First of all, once vertebrates colonize the land, oceans disappear from life's history, with one "exception" (documented in Figure 1) that actually illustrates the rule: If a "highly evolved" land creature returns to the sea, it may be shown as a representative of diversity within a stage of progress. Thus, Mesozoic marine reptiles may be depicted as contemporaries of ruling dinosaurs on land, but fishes living at the same time are invisible because their stage has been superseded in evolution's upward march. Tertiary whales are in because mammals then rule the land, but both marine reptiles and fishes of the same period are out as bypassed forms.

Second, the sequence of land animals only displays our anthropocentric view of shifting mastery through time, not a fair record of changing diversity. Fishes are banished once amphibians and reptiles colonize the land—but why punish fishes for what a few odd relatives did in disparate and unknown environments, especially when oceans, continuously dominated by fishes among vertebrates, cover some 70 percent of the earth's surface? The origin of mammals extirpates all amphibians and reptiles from view, even though they continue to flourish and to influence mammalian life in ways ranging from Mosaic plagues to the temptation of Eve. The last few paintings always depict humans, even though we are but one species in a small group of mammals (the order Primates contains about two hundred species among four thousand or so for all mammals), while the greatest successes of mammalian evolution—bats, rats, and antelopes—remain invisible.

Let me not carp unfairly. If these pageantries only claimed to be illustrating the ancestry of our tiny human twig on life's tree, then I would not complain, for I cannot quarrel unduly with such a parochial decision, stated up front. But these iconographic sequences always purport to be illustrating *the* history of life, not a tale of a twig. Consider the titles of the three series partly depicted in Figure 1: "The earth before the flood," "The parade of life through the ages," and "Prehistoric animals." An analogy might help in illustrating the oddity of such a pageant: Suppose that we wanted to stage a parade illustrating the growth of America's coterminous forty-eight states through time. Would we let the float for New England ride only for the first mile, and then withdraw it permanently from view? Would we then add the Northwest Territories, the Louisiana Purchase, and the western lands in sequence, permitting only one float at a time by dismantling the preceding float after each new introduction? Would we be adequately showing the apotheosis of American expansion if the parade ended with a single float celebrating that little sliver of the southwest known as the Gadsden Purchase?

Similarly, much as we may love ourselves, *Homo sapiens* is not representative, or symbolic, of life as a whole. We are not surrogates for arthropods (more than 80 percent of animal species), or exemplars of anything either particular or typical. We are the possessors of one extraordinary evolutionary invention called consciousness—the factor that permits us, rather than any other species, to ruminate about such matters (or, rather, cows ruminate and we cogitate). But how can this invention be viewed as the distillation of life's primary thrust or direction when 80 percent of multicellularity (the phylum Arthropoda) enjoys such evolutionary success and displays no trend to neurological complexity through time—and when our own neural elaboration may just as well end up destroying us as sparking a move to any other state that we would choose to designate as "higher"?

Why, then, do we continually portray this pitifully limited picture of one little stream in vertebrate life as a model for the whole multicellular pageant? Yet how many of us have ever looked at such a standard iconographic sequence and raised any question about its basic veracity? The usual iconography seems so right, so factual. I shall argue in this book that our unquestioning approbation of such a scheme provides our culture's most prominent example of a more extensive fallacy in reasoning about trends—a focus on particulars or abstractions (often biased examples like the lineage of *Homo sapiens*), egregiously selected from a totality because we perceive these limited and uncharacteristic examples as moving somewhere—when we should be studying variation *in the entire system* (the "full house" of my title) and its changing pattern of spread through time. I will emphasize the set of trends that inspires our greatest interest—supposed improvements through time. And I shall illustrate an unconventional mode of interpretation that seems obvious once stated, but rarely enters our mental framework—trends properly viewed as results of expanding or contracting variation, rather than concrete entities moving in a definite direction. This book, in other words, treats the *"spread* of *excellence,"* or trends to improvement best interpreted as expanding or contracting variation.

· 2 ·

Darwin Amidst the Spin Doctors

Biting the Fourth Freudian Bullet

I have often had occasion to quote Freud's incisive, almost rueful, observation that all major revolutions in the history of science have as their common theme, amidst such diversity, the successive dethronement of human arrogance from one pillar after another of our previous cosmic assurance. Freud mentions three such incidents: We once thought that we lived on the central body of a limited universe until Copernicus, Galileo, and Newton identified the earth as a tiny satellite to a marginal star. We then comforted ourselves by imagining that God had nevertheless chosen this peripheral location for creating a unique organism in His image—until Darwin came along and "relegated us to descent from an animal world." We then sought solace in our rational minds until, as Freud notes in one of the least modest statements of intellectual history, psychology discovered the unconscious.

Freud's statement is acute, but he left out several important revolutions in the pedestal-smashing mode (I offer no criticism of Freud's insight here, for he tried only to illustrate the process, not to provide an exhaustive list). In particular, he omitted the major contribution made to this sequence by my own field of geology and paleontology—the temporal counterpart to Copernicus's spatial discoveries. The biblical story, read literally, was so comforting: an earth only a few thousand years old, and occupied for all but the first five days by humans as dominant living creatures. The history of the earth becomes coextensive with the story of human life. Why not, then, interpret the physical universe as existing for and because of us?

But paleontologists then discovered "deep time," in John McPhee's felicitous phrase. The earth is billions of years old, receding as far into time as the visible universe extends into space. Time itself poses no Freudian threat, for if human history had occupied all these billions, then we might have increased our arrogance by longer hegemony over the planet. The Freudian dethronement occurred when paleontologists revealed that human existence only fills the last micromoment of planetary time—an inch or two of the cosmic mile, a minute or two in the cosmic year. This phenomenal restriction of human time posed an obvious threat, especially in conjunction with Freud's second, or Darwinian, revolution. For such a limitation has a "plain meaning"—and plain meanings are usually correct (even though many of our most fascinating intellectual revolutions celebrate the defeat of apparently obvious interpretations): If we are but a tiny twig on the floridly arborescent bush of life, and if our twig branched off just a geological moment ago, then perhaps we are not a predictable result of an inherently progressive process (the vaunted trend to progress in life's history); perhaps we are, whatever our glories and accomplishments, a momentary cosmic accident that would never arise again if the tree of life could be replanted from seed and regrown under similar conditions.

In fact, I would argue that all these "plain meanings" are true, and that we should revel in our newfound status and attendant need to construct meanings by and for ourselves—but this is another story for another time. I called this other story *Wonderful Life* (Gould, 1989). The theme for the present book, something of a philosophical "companion volume," is *Full*

House. For now, I only point out that this plain meaning is profoundly antithetical to some of the deepest social beliefs and psychological comforts of Western life—and that popular culture has therefore been unwilling to bite this fourth Freudian bullet.

Only two options seem logically available in our attempted denial. We might, first of all, continue to espouse biblical literalism and insist that the earth is but a few thousand years old, with humans created by God just a few days after the inception of planetary time. But such mythology is not an option for thinking people, who must respect the basic factuality of both time's immensity and evolution's veracity. We have therefore fallen back upon a second mode of special pleading—Darwin among the spin doctors. How can we tell the story of evolution with a slant that can validate traditional human arrogance?

If we wish both to admit the restriction of human time to the last micromoment of planetary time, and to continue our traditional support for our own cosmic importance, then we have to put a spin on the tale of evolution. I believe that such a spin would seem ridiculous *prima facie* to the metaphorical creature so often invoked in literary works to symbolize utter objectivity—the dispassionate and intelligent visitor from Mars who arrives to observe our planet for the first time, and comes freighted with no *a priori* expectations about earthly life. Yet we have been caught in this particular spin so long and so deeply that we do not grasp the patent absurdity of our traditional argument.

This positive spin rests upon the fallacy that evolution embodies a fundamental trend or thrust leading to a primary and defining result, one feature that stands out above all else as an epitome of life's history. That crucial feature, of course, is progress—operationally defined in many different ways[1] as a tendency for life to increase in anatomical complexity, or

---

1. One basically sophistic argument against progress holds that the word itself is too vague or subjective, and that the concept should be dropped for lack of rigor in description. This argument is a cop-out, and I will certainly not invoke such a lame defense in this book. Progress is too vague to stand by itself, but a variety of operational surrogates have been proposed—ranging from something as precise and measurable as brain size to more general, but still definable, notions as anatomical complexity (usually construed as number of parts and their degree of differentiation, assessed in various ways). I shall argue that progress as the primary thrust of life's history cannot be defended even for these operational surrogates.

neurological elaboration, or size and flexibility of behavioral repertoire, or any criterion obviously concocted (if we would only be honest and introspective enough about our motives) to place *Homo sapiens* atop a supposed heap.

We might canvass a range of historians, psychologists, theologians, and sociologists for their own distinctive views on why we feel such a need to validate our existence as a predictable cosmic preference. I can speak only from my own perspective as a paleontologist in the light of the fourth Freudian revolution: We are driven to view evolution's thrust as predictable and progressive in order to place a positive spin upon geology's most frightening fact—the restriction of human existence to the last sliver of earthly time. With such a spin, our limited time no longer threatens our universal importance. We may have occupied only the most recent moment as *Homo sapiens,* but if several billion preceding years displayed an overarching trend that sensibly culminated in our mental evolution, then our eventual origin has been implicit from the beginning of time. In one important sense, we have been around from the start. *In principio erat verbum.*

We may easily designate belief in progress as a potential bias, but some biases are true: my utterly subjective rooting preferences led me to love the Yankees during the 1950s, but they were also, objectively, the best team in baseball. Why should we suspect that progress, as the defining thrust of life's history, is not true? After all, and quite apart from our wishes, doesn't life manifestly become more complex? How can such a trend be denied in the light of paleontology's most salient fact: In the beginning, 3.5 billion years ago, all living organisms were single cells of the simplest sort, bacteria and their cousins; now we have dung beetles, seahorses, petunias, and people. You would have to be a particularly refractory curmudgeon, one of those annoying characters who loves verbal trickery and empty argument for its own sake, to deny the obvious statement that progress stands out as the major pattern of life's history.

This book tries to show that progress is, nonetheless, a delusion based on social prejudice and psychological hope engendered by our unwillingness to accept the plain (and true) meaning of the fourth Freudian revolution. I shall not make my case by denying the basic fact just presented:

Long ago, only bacteria populated the earth; now, a much broader diversity includes *Homo sapiens*. I shall argue instead that we have been thinking about this basic fact in a prejudiced and unfruitful way—and that a radically different approach to trends, one that requires a revision of even more basic mental habits dating at least to Plato, offers a more profitable framework. This new vantage point will also help us to understand a wide range of puzzling issues from the disappearance of 0.400 hitting in baseball to the absence of modern Mozarts and Beethovens.

Can We Finally Complete Darwin's Revolution?

The bias of progress expresses itself in various ways, from naive versions of pop culture to sophisticated accounts in the most technical publications. I do not, of course, claim that all, or even many, people accept the maximally simplistic account of a single ladder, with humans on top—although this imagery remains widespread, even in professional journals. Most writers who have studied some evolutionary biology understand that evolution is a copiously branching bush with innumerable present outcomes, not a highway or a ladder with one summit. They therefore recognize that progress must be construed as a broad, overall, average tendency (with many stable lineages "failing" to get the "message" and retaining fairly simple form through the ages).

Nonetheless, however presented, and however much the sillier versions may be satirized and ridiculed, claims and metaphors about evolution as progress continue to dominate all our literatures—a testimony to the strength of this primary bias. I present a few items, almost randomly selected from my burgeoning files:

• From *Sports Illustrated,* August 6, 1990, Denver Broncos veteran Karl Mecklenburg, on being shifted from defensive end to inside linebacker to a new position as outside linebacker: "I'm moving right up the evolutionary ladder."

- From a correspondent, writing from Maine on January 18, 1987, and puzzled because he cannot spot the fallacy in a creationist tract: The pamphlet "shows that well dated finds of many species of man show no advancement within a species over the thousands of years the species existed. Also many species appear to have existed concurrently. Both these finds contradict the precepts of evolution which insists each species advances towards the next higher."

- From another correspondent, in New Jersey (December 22, 1992), a professional scientist this time, expressing his understanding that life *as a totality*, not just selected lineages at pinnacles of their groups, should progress through time: "I assume that as evolution proceeds, a greater and greater degree of specialization occurs with regard to structure and physiological activity. After a billion years or more of biological evolution I would think that the extant species are relatively highly specialized."

- From a correspondent in England on June 16, 1992, really putting it on the line: "Life has a sort of 'built-in' drive towards complexity, matched by no drive to de-complexity. . . . Human consciousness was inevitable once things got started on Complexity Road in the first place."

- From a leading high school biology textbook, published in 1966, and providing a classic example of a false inference (the first sentence) drawn from a genuine fact (the second sentence): "Most descriptions of the pattern of evolution depend upon the assumption that organisms tend to become more and more complicated as they evolve. If this assumption is correct, there would have been a time in the past when the earth was inhabited only by simple organisms."

- From America's leading professional journal, *Science,* in July 1993: An article titled "Tracing the Immune System's Evolutionary History" rests upon the peculiar premise, intelligible only if "everybody knows" about life's progress through time, that we should be surprised to discover sophisticated immune devices in "the lower organisms" (their phrase, not mine). The article claims to be reporting a remarkable insight: "the immune system in simpler organisms isn't just a less sophisticated version of our own." (Why should anyone have ever held such a view of "others" as basically "less than us," especially when the "simpler organisms" under discussion are arthropods with 500 million years of evolutionary separation from vertebrates, and when all scientists recognize the remarkable diversity and complexity of chemical defense systems maintained by many insects?) The article also expresses surprise that "creatures as far down the evolutionary ladder as sponges can recognize tissue from other species." If our leading professional journal still uses such imagery about evolutionary ladders, why should we laugh at Mr. Mecklenburg for his identical metaphor?

The allure of this conventional imagery is so great that I have fallen into the trap myself—by presenting my examples as an ascending ladder from the central pop icon of a sports hero, through letters of increasing sophistication, to textbooks, to an article in *Science.* Yet the last shall be first, and my linear sequence bends into a circle of error, as both my initial and final examples misuse the identical phrase about an "evolutionary ladder." At least the linebacker was trying to be funny!

These lists of error could go on forever, but let me close this section with two striking examples representing the pinnacle (there we go with progress metaphors again) of fame and achievement in the domains of popular and professional life.

• Popular culture's leading version: Psychologist M. Scott Peck's *The Road Less Traveled,* first published in 1978, must be the greatest success in the history of our distinctive and immensely popular genre of "how-to" treatises on personal growth. This book has been on the *New York Times* best-seller list for more than six hundred weeks, placing itself so far in first place for total sales that we need not contemplate any challenge in our lifetime. Peck's book includes a section titled "The Miracle of Evolution" (pages 263–68).

Peck begins his discussion with a classic misunderstanding of the second law of thermodynamics:

The most striking feature of the process of physical evolution is that it is a miracle. Given what we understand
of the universe, evolution should not occur; the phenomenon should not exist at all. One of the basic natural laws
is the second law of thermodynamics, which states that
energy naturally flows from a state of greater organization to a state of lesser organization. . . . In other words,
the universe is in a process of winding down.

But this statement of the second law, usually portrayed as increase of
entropy (or disorder) through time, applies only to closed systems that receive no inputs of new energy from exterior sources. The earth is not a
closed system; our planet is continually bathed by massive influxes of solar
energy, and earthly order may therefore increase without violating any natural law. (The solar system as a whole may be construed as closed and
therefore subject to the second law. Disorder does increase in the entire
system as the sun uses up fuel, and will ultimately explode. But this final
fate does not preclude a long and local buildup of order in that little corner of totality called the earth.)

Peck designates evolution as miraculous for violating the second law
in displaying a primary thrust toward progress through time:

The process of evolution has been a development of organisms from lower to higher and higher states of complexity, differentiation, and organization. . . . [Peck then
writes, in turn, about a virus, a bacterium, a paramecium,
a sponge, an insect, and a fish—as if this motley order represented an evolutionary sequence. He continues:] And
so it goes, up the scale of evolution, a scale of increasing
complexity and organization and differentiation, with
man who possesses an enormous cerebral cortex and ex-
traordinarily complex behavior patterns, being, as far as
we can tell, at the top. I state that the process of evolution
is a miracle, because insofar as it is a process of increasing organization and differentiation it runs counter to
natural law.

Peck then summarizes his view as a diagram (redrawn here as Figure 2), a stunning epitome of the grand error that the bias of progress imposes upon us. He recognizes the primary fact of nature that stands so strongly against any simplistic view of progress (and, as I shall show later in this book, debars the subtler versions as well)—rarity of the highest form (humans) versus ubiquity of the lowest (bacteria). If progress is so damned good, why don't we see more of it?

Peck tries to pry victory from the jaws of defeat by portraying life as thrusting upward against an entropic downward tug:

> The process of evolution can be diagrammed by a pyramid, with man, the most complex but least numerous organism, at the apex, and viruses, the most numerous but least complex organisms, at the base. The apex is thrusting out, up, forward against the force of entropy. Inside the pyramid I have placed an arrow to symbolize this thrusting evolutionary force, the "something" that has so successfully and consistently defied "natural law" over millions upon millions of generations and that must itself represent natural law as yet undefined.

Note how this simple diagram encompasses all the major errors of progressivist bias. First, although Peck supposedly rejects the most naive version of life's ladder, he places an explicit linear array right under his apex of progress as the motor of upward thrusting. Two features of this reintroduced ladder reveal Peck's lack of attention and sympathy for natural history and life's diversity. I am, I confess, galled by the insouciant sweep that places only "colonial organisms" into the enormous domain between bacteria and vertebrates—where they must stand for all eukaryotic unicellular organisms and all multicellular invertebrates as well, though neither category includes many colonial creatures! But I am equally chagrined by Peck's names for the prehuman vertebrate sequence: fish, birds, and animals. I know that fish gotta swim and birds gotta fly, but I certainly thought that they, and not only mammals, were called animals.

Second, the model of life's upward thrust versus inorganic nature's downward tug allows Peck to view progress as evolution's most power-

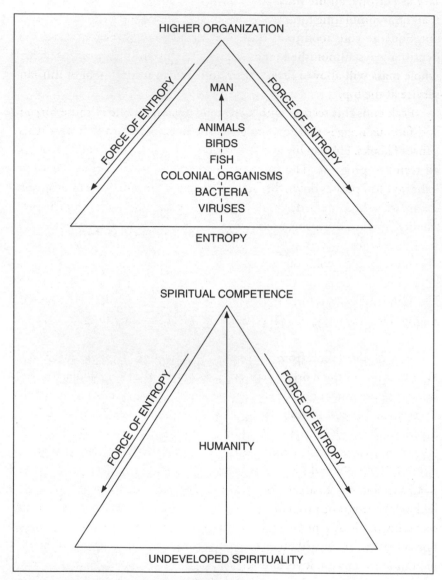

FIGURE 2

Two biased views of evolution as progress from M. Scott Peck's best-selling *The Road Less Traveled*. Above, the supposed pyramid of life's upwardly driving complexity. Below, the same scheme applied to the supposed development of human spiritual competence.

ful and universal trend, even against the observation that most organisms don't get very far along the preferred path: against so powerful an adversary as entropy, all life must stand and shove together from the base, so that the accumulating force will push a favored few right up to the top and out. Squeeze your toothpaste tube *from the bottom,* just as Mom and the dentist always admonished (and so few of us do), and the pressure of the whole mass will allow a little stream to reach an utmost goal of human service at the top.

Peck ends this section with a crescendo based on one of those forced and fatuous images that sets my generally negative attitude toward this genre of books. Human life and striving become a microcosm of life's overall trend to progress. The force of entropy (also identified as our own lethargy) still pushes down, but love, standing in for the drive of progress (or are they the same?), drives us from the low state of "undeveloped spirituality" toward the acme, or pyramidal point, of "spiritual competence." Peck concludes by writing, "Love, the extension of the self, is the very act of evolution. It is evolution in progress. The evolutionary force, present in all of life, manifests itself in mankind as human love. Among humanity love is the miraculous force that defies the natural law of entropy." Sounds mighty nice and cozy, but I'll be damned if it means anything.

• A similar vision from the professional heights. My colleague E. O. Wilson is one of the world's greatest natural historians. If anyone understands the meaning and status of species and their interrelationships, this unparalleled expert on ants, and tireless crusader for preservation of biodiversity, should be the paragon. I enjoyed his book *The Diversity of Life* (1992), and reviewed it favorably in the leading British journal *Nature* (Gould, 1993). Ed and I have our disagreements about a variety of issues, from sociobiology to arcana of Darwinian theory, but we ought to be allied on the myth of progress, if only because success in our profession's common battle for preserving biodiversity requires a reorientation of human attitudes toward other species—from little care and maximal exploitation to interest, love, and respect. How can this change occur if we continue to view ourselves as better than all others by cosmic design?

Nonetheless, Wilson uses the oldest imagery of the progressivist view in epitomizing the direction of life's history as a series of formal Ages (with uppercase letters, no less)—a system used by virtually all popular works and textbooks in my youth, but largely abandoned (I thought), for reform so often affects language first (as in our eternal debates about political correctness and the proper names for groups and genders), and concepts only later:

> They [arthropods as the first land animals] were followed by the amphibians, evolved from lobe-finned fishes, and a burst of land vertebrates, relative giants among land animals, to inaugurate the Age of Reptiles. Next came the Age of Mammals and finally the Age of Man.

These words do not represent a rhetorical slip into comfortable, if antiquated, phraseology, for Wilson also provides his explicit defense of progress, ending with a line that I found almost chilling:

> Many reversals have occurred along the way, but the overall average across the history of life has moved from the simple and few to the more complex and numerous. During the past billion years, animals as a whole evolved upward in body size, feeding and defensive techniques, brain and behavioral complexity, social organization, and precision of environmental control. . . . Progress, then, is a property of the evolution of life as a whole by almost any conceivable intuitive standard, including the acquisition of goals and intentions in the behavior of animals. It makes little sense to judge it irrelevant. Attentive to the adjuration of C. S. Peirce, let us not pretend to deny in our philosophy what we know in our hearts to be true.

Peirce may have been our greatest thinker, but his line in this context almost sounds scary. Nothing could be more antithetical to intellectual reform than an appeal *against* thoughtful scrutiny of our most hidebound mental habits—notions so "obviously" true that we stopped thinking about them generations ago, and moved them into our hearts and bosoms. Please do not forget that the sun really does rise in the east, move through the sky each day, and set in the west. What knowledge could be more visceral than the earth's central stability and the sun's subordinate motion?

Darwin was born on the same day as Lincoln, and "officially" inaugurated the revolution that bears his name when he published the *Origin of Species* in 1859. During the centennial celebrations in 1959, the great American geneticist H. J. Muller dampened festivities with an address titled "One Hundred Years Without Darwin Are Enough." Muller treated the revolution's failure to penetrate at two opposite ends of a spectrum—creationism's continuing hold over much of American pop culture, and limited understanding of natural selection among well-educated people content with the factuality of evolution.

But I think that something even larger, and standing in the middle of this spectrum, has always ranked as the greatest impediment to completing the Darwinian revolution. Freud was right in identifying suppression of human arrogance as the common achievement of great scientific revolutions. Darwin's revolution—the acceptance of evolution with *all* major implications, the second blow in Freud's own series—has never been completed. In Freud's terms, the revolution will not be fulfilled when Mr. Gallup can find no more than a handful of deniers, or when most Americans can give an accurate epitome of natural selection. Darwin's revolution will be completed when we smash the pedestal of arrogance and own the plain implications of evolution for life's nonpredictable nondirectionality—and when we take Darwinian topology seriously, recognizing that *Homo sapiens,* to recite the revised litany one more time, is a tiny twig, born just yesterday on an enormously arborescent tree of life that would never produce the same set of branches if regrown from seed. We grasp at the straw of progress (a desiccated ideological twig) because we are still not ready for the Darwinian revolution. We crave progress as our best hope for retaining human arrogance in an evolutionary world. Only in these terms can I understand why such a poorly formulated and improbable argument maintains such a powerful hold over us today.

· 3 ·

Different Parsings,
Different Images of Trends

Fallacies in the Reading and Identification of Trends

The more important the subject and the closer it cuts to the bone of our hopes and needs, the more we are likely to err in establishing a framework for analysis. We are story-telling creatures, products of history ourselves. We are fascinated by trends, in part because they tell stories by the basic device of imparting directionality to time, in part because they so often supply a moral dimension to a sequence of events: a cause to bewail as something goes to pot, or to highlight as a rare beacon of hope.

But our strong desire to identify trends often leads us to detect a directionality that doesn't exist, or to infer causes that cannot be sustained.

The subject of trends has inspired and illustrated some of the classic fallacies in human reasoning. Most prominently, since people seem to be so bad at thinking about probability and so prone to read pattern into sequences of events, we often commit the fallacy of spotting a "sure" trend and speculating about causes, when we observe no more than a random string of happenings.

In the classic case, most people have little sense of how often an apparent pattern will emerge in purely random data. Take the standard illustration of coin flipping: we compute the probability of sequences by multiplying the chances of individual events. Since the probability for heads is always 1/2, the chance of flipping five heads in a row is $1/2 \times 1/2 \times 1/2 \times 1/2 \times 1/2$, or one in thirty-two—rare to be sure, but something that will happen every once in a while for no reason but randomness. Many people, however, particularly if they are betting on tails, will read five heads in a row as *prima facie* evidence of cheating. People have been shot and killed for less—in life as well as in Western movies.

In my favorite, more subtle example of the same error, T. Gilovich, R. Vallone, and A. Tversky debunked a phenomenon that every basketball fan and player absolutely "knows" to be true—"hot hands," or streaks of successive baskets, magic minutes of "getting into the groove" or "finding the range," when every shot hits. The phenomenon sounds so obvious: when you're hot you're hot, and when you're not you're not. But "hot hands" does not exist. My colleagues studied every basket made by the Philadelphia 76ers for more than a season. They made two debunking discoveries: first, the probability of making a second basket did not rise following a successful shot; second, and more important, the number of "runs," or successful baskets in sequence, did not exceed the predictions of a standard random, or coin-tossing, model. Remember that, on average, you will flip five heads in a row once in every thirty-two sequences of five tosses. We can, by analogy, compute expected runs for any basketball player. Suppose that Mr. Swish, a particularly good shooter, succeeds in 60 percent of his field-goal attempts. He should then notch six baskets in a row once every 20 sequences or so ($0.6 \times 0.6 \times 0.6 \times 0.6 \times 0.6 \times 0.6$, for 0.047, or 4.7 percent). If Swish's actual play includes sequences of six at about this rate, then we have no evidence for hot hands, but only for Swish playing in his characteristic manner for each shot independently. Gilovich,

Vallone, and Tversky found no sequences beyond the range of random expectations.

My colleague Ed Purcell, a Nobel Prize winner in physics but just a keen baseball fan in this context, then did a similar study of baseball streaks and slumps, and we published the results together (Gould, 1988). Purcell found that among all runs, the subject of so much mythology about heroes (and goats), only one record stands beyond reasonable probability, and should not have happened at all—Joe DiMaggio's fifty-six-game hitting streak in 1941—thus validating the feeling of many fans that DiMaggio's splendid run is the greatest achievement in modern sports (and exonerating all the poor schlumps whose runs of failure lie entirely within the expectations of their characteristic probabilities!).

As one final example, probably more intellectual energy has been invested in discovering (and exploiting) trends in the stock market than in any other subject—for the obvious reason that stakes are so high, as measured in the currency of our culture. The fact that no one has ever come close to finding a consistent way to beat the system—despite intense efforts by some of the smartest people in the world—probably indicates that such causal trends do not exist, and that the sequences are effectively random.

In the second most prominent fallacy about trends, people correctly identify a genuine directionality, but then fall into the error of assuming that something else moving in the same direction at the same time must be acting as the cause. This error, the conflation of correlation with causality, arises for the obvious reason (once you think about it) that, at any moment, oodles of things must be moving in the same direction (Halley's comet is receding from earth and my cat is getting more ornery)—and the vast majority of these correlated sequences cannot be causally related. In the classic illustration, a famous statistician once showed a precise correlation between arrests for public drunkenness and the number of Baptist preachers in nineteenth-century America. The correlation is real and intense, but we may assume that the two increases are causally unrelated, and that both arise as consequences of a single different factor: a marked general increase in the American population.

The error detailed in this book has not often been named or identified, but may be just as prominent in our fallacious thinking about trends.

I shall focus on two central examples from two dramatically different cultural realms: "Why does no one hit 0.400 anymore in baseball?" and "How does progress characterize the history of life?" These are classic trends, in the sense that each encapsulates the essence and history of an important institution, and both have moral implications—one, in baseball, apparently trying to tell us that something about modern life causes excellence, or old-fashioned virtue, to degenerate; the other, for life, providing our necessary solace and excuse for continuing to view ourselves as lords of all.

I shall not use the juxtaposition of these examples to present pap and nonsense about how life imitates baseball, or vice versa. But I will show that the same error has led us to view both trends the wrong way round. Straighten out the fallacy, and you will see that the disappearance of 0.400 hitting illustrates the increasing excellence of play in baseball (however paradoxical such a claim may sound at first)—while life, on the other hand, shows no general thrust to improvement, but just adds an occasional exemplar of complexity in the only region of available anatomical space, while maintaining, for more than 3 billion years, an unvarying bacterial mode. Baseball has improved, but life has always been, and will probably always remain until the sun explodes, in the Age of Bacteria.

The common error lies in failing to recognize that apparent trends can be generated as by-products, or side consequences, of expansions and contractions in the amount of variation within a system, and not by anything directly moving anywhere. Average values may, in fact, stay constant within the system (as average batting percentages have done in major-league baseball, and as the bacterial mode has remained for life)—while our (mis)perception of a trend may represent only our myopic focus on rare objects at one extreme in a system's variation (as this periphery expands or contracts). And the reasons for expansion or contraction of a periphery may be very different from causes for a change in average values. Thus, if we mistake the growth or shrinkage of an edge for movement of an entire mass, we may devise a backwards explanation. I will show that the disappearance of 0.400 hitting marks the shrinkage of such an edge caused by increasing excellence in play, not the extinction of a cherished entity (which would surely signify degeneration of something, and a *loss* of excellence).

Let me illustrate this unfamiliar concept with a simple (and silly) example to show how, in two cases, an apparent trend may arise only by expansion or contraction of variation. In both cases we tend to misinterpret a phenomenon because we maintain such strong preferences for viewing trends as entities moving somewhere.

The one hundred inhabitants of a mythical land subsist on an identical diet and all weigh one hundred pounds. In my first case, an argument about nutrition develops, with some folks pushing a new (and particularly calorific) brand of cake, and others advocating increased abstemiousness. Most members of the population don't give a damn and stay where they are, but ten folks eat copious amounts of cake and now average 150 pounds, while ten others run and starve to reach an average weight of fifty pounds. The mean of the population hasn't altered at all, remaining right at its old value of one hundred pounds—but variation in weight has expanded markedly (and symmetrically in both directions).

Cake-makers, pushing the aesthetic beauty of the new and fuller look, might celebrate a trend to greater weight by focusing on the small subset of people under their influence, and ignoring the others—just as the running-and-dieting moralists might exalt twigginess and praise a supposed trend in this direction by isolating their own small subset. But no general trend has occurred at all, at least in the usual sense. The average of the population has not altered by a single pound, and most people (80 percent in this case) have not varied their weight by an ounce. The only change has been a symmetrical expansion of variation on both sides of a constant mean weight. (You may recognize this increased spread as significant, of course, but we usually don't describe such nondirectional changes as "trends.")

You may choose to regard this example as both silly and transparent. Few of us would have any trouble identifying the actual changes, and we would laugh the shills of both cake-makers and runner-dieters out of town, if they tried to pass off the changes in their small subset as a general trend. But bear with me, for I shall show that many phenomena often perceived as trends, and either celebrated or lamented with gusto and acres of printer's ink—the disappearance of 0.400 hitting among them—also represent symmetrical changes of variation around constant mean values, and therefore display the same fallacy, though better hidden.

My second case features a totalitarian society ruled by the runner-dieters. They have been pushing their line for so long that everyone has succumbed to social pressure and weighs fifty pounds. A more liberal regime takes over and permits free discussion about ideal weights. Fine, but for one catch imposed by physiology rather than politics: fifty pounds is the lower limit for sustaining life, and no one can get any thinner. Therefore, although citizens are now free to alter their weight, only one direction of change is possible. The great majority of inhabitants remain content with the old ways and elect to maintain themselves at fifty pounds. Fifteen percent of the population revels in its newfound freedom and begins to gain weight with abandon. Six months later, these fifteen individuals average seventy-five pounds; after a year, one hundred pounds; and after two years, 150 pounds.

The statistical spin doctors for the fat fifteen now step in. They argue that their clients' point of view is sweeping through the whole society, as clearly indicated by the steady increase of mean weight for the entire population. And who can deny their evidence? They even present a fancy graph (shown here as Figure 3). Before the liberation, average weight stood at fifty pounds; after six months the mean rises to 53.8 pounds (the average for eighty-five remaining at fifty pounds, and fifteen rising to seventy-five pounds); after a year to 57.5 pounds; and after two years to sixty-five pounds (an increase of 30 percent from the original fifty)—a steady, unreversed, and substantial rise.

Again, you may view this example as silly (and purposely chosen to illustrate the obvious nature of the point, once you understand the whole system and its variation). Few people would be fooled, so long as they grasped the totality of the story, and knew that most members of the population had not changed their weight, and that the steady increase in mean values arises as an artifact produced by amalgamating two entirely different subpopulations—a majority of stalwarts with a minority of revolutionaries. But suppose you didn't appreciate the whole tale, and only listened to the statistical spin doctors for the fat fifteen. Suppose, in addition, that you tended to imbue mean values (as I fear most of us do) with a reality transcending actual individuals and the variation among them. You might then be persuaded from Figure 3 that a general trend has swept through the population, thrusting it *as a whole* toward greater average weights.

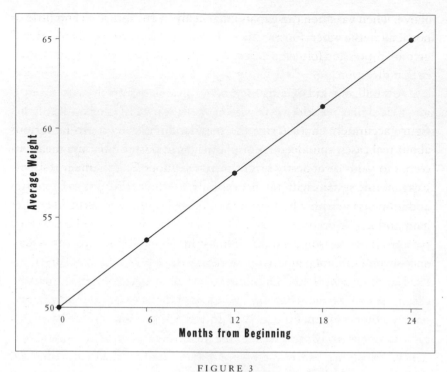

FIGURE 3

Average weight of my hypothetical population plotted against time to show how a false impression of an overall trend may be generated.

We are more likely to be fooled by the second case, where limits to variation on one side of the average permit change in only one direction. The rise of mean values isn't "false" in this second case, but the supposed trend is surely misleading in the sense of Mark Twain's or Disraeli's famous line (the quote has been attributed to both) about three kinds of falsification—"lies, damned lies, and statistics." I will present the technicalities later, but let me quickly state why such false impressions can emerge from correct data in this case—as so often exploited by economic pundits and political spin doctors. As in the cliché about skinning cats, there is more than one way to represent an "average." The most common method, technically called the *mean,* instructs us to add up all the values and divide by the number of cases. If ten kids have ten dollars among them, the mean wealth per kid is one dollar. But means can be grossly misleading—and never more so than in the type of example purposely chosen above: when variation can expand markedly in one direction and little or not at all in the other. For means will then drift toward the open end and give an impression (often quite false) that the whole population has moved in that direction.

After all, one kid may have a ten-dollar bill, and the other nine nothing. One dollar per kid would still be the mean value, but would such a figure accurately characterize the population? Similarly, to be serious about real cases, spin doctors for politicians in power often use mean incomes to paint dishonestly bright pictures. Suppose that, under a super-Reaganomic system with tax breaks only for the rich, a few millionaires add immense wealth while a vast mass of people at the poverty line either gain nothing or become poorer. The mean income may rise because one tycoon's increase from, say, $6 million to $600 million per year may balance several million paupers. If one man gains $594 million and one hundred million people lose five dollars each (for a total of $500 million), mean income for the whole population will still rise—but no one would dare say (honestly) that the average person was making more money.

Statisticians have developed other measure of average, or "central tendency," to deal with such cases. One alternative, called the *mode,* is defined as the most common value in the population. No mathematical rule can tell us which measure of central tendency will be most appropriate for any particular problem. Proper decisions rest upon knowledge of all factors in a given case, and upon basic honesty.

Would anyone dispute a claim that modes, rather than means, provide a better understanding of all the examples presented above? The modal amount of money for the ten kids is zip. The modal income for our population remains constant (or falls slightly), while the mean rises because one tycoon makes an immense killing. The modal weight for the population of my second silly example remains at fifty pounds. The fifteen gainers increase steadily (and the mean of the whole population therefore rises), but who would deny that stability of the majority best characterizes the population as a whole? (At the very least, allow me that you cannot represent the population by the rising mean values of Figure 3 if, for whatever personal reason, you choose to focus on the gainers—and that you must identify the stability of the majority as a major phenomenon.) I belabor this point because my second focal example, progress in the history

of life, emerges as a delusion on precisely the same grounds. A few creatures have evolved greater complexity in the only direction open to variation. The mode has remained rock-solid on bacteria throughout the
history of life—and bacteria, by any reasonable criterion, were in the beginning, are now, and ever shall be the most successful organisms on earth.

Variation as Universal Reality

I have tried to show how an apparent trend in a whole system—traditionally read as a "thing" (the population's average, for example) moving
somewhere—can represent a false reading based only on expansion or contraction of variation within the system. We make such errors either because we focus myopically upon the small set of changing extreme values
and falsely read their alteration as a trend in the whole system (my first
case, to be illustrated by 0.400 hitting in baseball)—or because variation
can expand or contract in only one direction, and we falsely characterize
the system by a changing mean value, while a stable mode suggests a radically different interpretation (my second case, to be illustrated by the
chimera of progress as the primary thrust of life's history).

I am not saying that all trends fall victim to this error (genuine "things"
do move somewhere sometimes), or that this "fallacy of reified variation"[2]
exceeds in importance the two more commonly recognized errors of confusing trends with random sequences, or conflating correlation with

---

2. *Reification* is an unfamiliar word, but this term describes the fallacy so well that I hasten
to use (and explain) it. As coined by philosophers and social scientists in the mid-nineteenth
century, *reification* refers to "the mental conversion of a person or abstract concept into a
thing" *(Oxford English Dictionary)*. The word comes from the Latin *res,* meaning "thing" (a
republic, or *res publica,* is the people's thing). When committing the error discussed in
this book, we abstract the variation within a system into some measure of central tendency,
like the mean value—and then make the mistake of reifying this abstraction and interpreting the mean as a concrete "thing"; we then compound our error by assuming that changes
in the mean must, *ipso facto,* be read as an entity moving somewhere. Or, in another version
of the same fallacy, we focus on extremes in variation and falsely reify these values as separate things, rather than treating them as an inextricable part of the entire system's variation.

causality. But the variational fallacy has caused us to read some of our most important, and most intensely discussed, cultural trends in an ass-backwards manner. I am also intrigued by this fallacy because our general misunderstanding or undervaluation of variation raises a much deeper issue about the basic perception of physical reality.

We often portray taxonomy as the dullest of all fields, as expressed in a variety of deprecatory metaphors: hanging garments on nature's coat-rack; placing items into pigeonholes; or (in an image properly resented by philatelists) sticking stamps into the album of reality. All these images clip the wings of taxonomy and reduce the science of classification to the dullest task of keeping things neat and tidy. But these portrayals also reflect a cardinal fallacy: the assumption of a fully objective nature "out there" and visible in the same way to any unprejudiced observer (the same image that I criticized in the first section of this chapter as "Huxley's chessboard"). If such a vision could be sustained, I suppose that taxonomy would become the most boring of all sciences, for nature would then present a set of obvious pigeonholes, and taxonomists would search for occupants and shove them in—an enterprise requiring diligence, perhaps, but not much creativity or imagination.

But classifications are not passive ordering devices in a world objectively divided into obvious categories. Taxonomies are human decisions imposed upon nature—theories about the causes of nature's order. The chronicle of historical changes in classification provides our finest insight into conceptual revolutions in human thought. Objective nature does exist, but we can converse with her only through the structure of our taxonomic systems.

We may grant this general point, but still hold that certain fundamental categories present so little ambiguity that basic divisions must be invariant across time and culture. Not so—not for these, or for any subjects. Categories are human impositions upon nature (though nature's factuality offers hints and suggestions in return). Consider, as an example, the "obvious" division of humans into two sexes.

We may view male versus female as a permanent dichotomy, as expressions of two alternative pathways in embryological development and later growth. How else could we possibly classify people? Yet this "two-sex model" has only recently prevailed in Western history (see Laqueur,

1990; Gould, 1991), and could not hold sway until the mechanical philosophy of Newton and Descartes vanquished the Neoplatonism of previous worldviews. From classical times to the Renaissance, a "one-sex model" was favored, with human bodies ranged on a continuum of excellence, from low earthiness to high idealization. To be sure, people might clump into two major groups, called male and female, along this line, but only one ideal or archetypal body existed, and all actual expressions (real persons) had to occupy a station along a single continuum of metaphysical advance. This older system is surely as sexist as the later "two-sex model" (which posits innate and predetermined differences of worth from the start), but for different reasons—and we need to understand this history of radically altered taxonomy if we wish to grasp the depth of oppression through the ages. (In the "one-sex model," conventional maleness, by virtue of more heat, stood near the apex of the single sequence, while the characteristic female form, through relative weakness of the same generating forces, ranked far down the single ladder.)

This book treats the even more fundamental taxonomic issue of what we designate as a thing or an object in the first place. I will argue that we are still suffering from a legacy as old as Plato, a tendency to abstract a single ideal or average as the "essence" of a system, and to devalue or ignore variation among the individuals that constitute the full population. (Just consider our continuing hang-ups about "normality." When I was a new father, my wife and I bought a wonderful book by the famous pediatrician T. Berry Brazelton. He wrote to combat every parent's excessive fear that one standard of normality exists for a child's growth, and that anything your particular baby does must be judged against this unforgiving protocol. Brazelton used the simple device of designating three perfectly fine pathways, each exemplified by a particular child—one hellion, one in the middle, and one shy baby who, in gentle euphemism, was labeled "slow to warm up." Even three, instead of one, doesn't capture the richness of normal *variation*, but what a fine start in the right direction.)

In his celebrated analogy of the cave, Plato (in the *Republic*) held that actual organisms are only shadows on the cave's wall (empirical nature)—and that an ideal realm of essences must exist to cast the shadows. Few of us would maintain such an unbridled Platonism today, but we have never put aside this distinctive view that populations of actual individuals form a set of accidents, a collection of flawed examples, each necessarily imperfect and capable only of approaching the ideal to a certain extent. One might survey this pool of accidents and form some idea of the essence by cobbling together the best parts—the most symmetrical nose from this person, the most oval eyes from a second, the roundest navel from a third, and the best-proportioned toe from a fourth—but no actual individual can stand for the category's deeper reality.

Only by acknowledging this lingering Platonism can I understand the fatal inversion that we so often apply to calculated averages. In Darwin's post-Platonic world, variation stands as the fundamental reality and calculated averages become abstractions. But we continue to favor the older and opposite view, and to regard variation as a pool of inconsequential happenstances, valuable largely because we can use the spread to calculate an average, which we may then regard as a best approach to an essence. Only as Plato's legacy can I grasp the common errors about trends that make this book necessary: our misreading of expanding or contracting variation within a system as an average (or extreme) value moving somewhere.

I spoke in chapter 2 about completing Darwin's revolution. This intellectual upheaval included many components—in part (and already accomplished among educated people during Darwin's lifetime), the simple acceptance of evolution as an alternative to divine creation; in part (and still unfulfilled), Freud's pedestal-smashing recognition of *Homo sapiens* as only a recent twiglet on an ancient and enormous genealogical bush. But, in an even more fundamental sense, Darwin's revolution should be epitomized as the substitution of variation for essence as the central category of natural reality (see Mayr, 1963, our greatest living evolutionist, for a stirring defense of the notion that "population thinking," as a replacement for Platonic essentialism, forms the centerpiece of Darwin's revolution). What can be more discombobulating than a full inversion, or "grand flip," in our concept of reality: in Plato's world, variation is accidental, while essences record a higher reality; in Darwin's reversal, we value variation as a defining (and concrete earthly) reality, while averages (our closest operational approach to "essences") become mental abstractions.

Darwin knew that he was overturning fundamental ideas with venerable Greek ancestry. During his late twenties, in a youthful notebook about evolution, he wrote a wonderful, sardonic commentary about Plato's theory of essences—noting succinctly that the existence of innate ideas need not imply an ethereal realm of unchanging essential concepts, but may only indicate our descent from a material ancestor: "Plato says in *Phaedo* that our 'imaginary ideas' arise from the preexistence of the soul, are not derivable from experience—read monkeys for preexistence."

In his poem *History,* Ralph Waldo Emerson records the grand legacies held by this greatest of all subjects:

> *I am the owner of the sphere . . .*
> *Of Caesar's hand, and Plato's brain,*
> *Of Lord Christ's heart, and Shakespeare's strain.*

These legacies are our joy and inspiration, but also our weights and impediments. Read monkeys for preexistence, and read variation as the primary expression of natural reality.

Part Two

· · ·

Death and Horses: Two Cases for the Primacy of Variation

Before presenting my central examples of baseball and life, I offer two cases to illustrate my contention that our culture encodes a strong bias either to neglect or ignore variation. We tend to focus instead on measures of central tendency, and as a result we make some terrible mistakes, often with considerable practical import.

• • •

· 4 ·

Case One: A Personal Story

**Where any measure of central tendency
acts as a harmful abstraction,
and variation stands out
as the only meaningful reality**

In 1982, at age forty, I was diagnosed with abdominal mesothelioma, a rare and "invariably fatal" form of cancer (to cite all official judgments at the time). I was treated and cured by courageous doctors using an experimental method that can now save some patients who discover the disease in an early stage.

The cancer survivors' movement has spawned an enormous literature of personal testimony and self-help. I value these books, and learned much from them during my own ordeal. Yet, although I am a writer by trade, and although no experience could possibly be more intense than a long fight against a painful and supposedly incurable disease, I have never felt any urge or need to describe my personal experiences in prose. Instead, as an intensely private person, I view such an enterprise with horror. In all the years then and since, I have been moved to write only one short article about this cardinal portion of my life.

I accept and try to follow the important moral imperative that blessings must be returned with efforts of potential use to others. I am therefore enormously grateful that this article has been of value to people, and that so many readers have requested copies for themselves, or for a friend with cancer. But I did not write my article either from compulsion (as a personal testimony) or from obligation (to the moral requirement cited above). I wrote my piece, *The Median Is Not the Message,* from a different sort of intellectual need. I believe that the fallacy of reified variation—or failure to consider the "full house" of all cases—plunges us into serious error again and again: my battle against cancer had begun with a fine example of practical benefits to be gained by avoiding such an error, and I could not resist an urge to share the yarn.

We have come a long way from the bad old days, when cancer diagnoses were scrupulously hidden from most patients—both for the lamentable reason that many doctors regarded deception as a preferred pathway for maintaining control, and on the compassionate (if misguided) assumption that most people could not tolerate a word that conveyed ultimate horror and a sentence of death. But we cannot overcome obstacles with ignorance: consider what Franklin D. Roosevelt could have contributed to our understanding of disability if he had not hidden his paralysis with such cunning care, but had announced instead that he did not govern with his legs.

American doctors, particularly in intellectual centers like Boston, now follow what I regard as the best strategy for this most difficult subject: any information, no matter how brutal, will be given upon request (as compassionately and diplomatically as possible, of course); if you don't want to know, don't ask. My own doctor made only one departure from this sensible rule—and I forgave her immediately as soon as I faced the context. At our first meeting, after my initial surgery, I asked her what I could read to learn more about mesothelioma (for I had never heard of the disease). She replied that the literature contained nothing worth pursuing. But trying to keep an intellectual from books is about as effective as that old saw about ordering someone not to think about a rhinoceros. As soon as I could walk, I staggered over to the medical school library and punched *mesothelioma* into the computer search program. Half an hour later, surrounded by the latest articles, I understood why my doctor had erred on the side of limited information.

All the literature contained the same brutal message: mesothelioma is incurable, with a median mortality of eight months following diagnosis. A hot topic of late, expressed most notably in Bernie Siegel's best-selling books, has emphasized the role of positive attitude in combating such serious diseases as cancer. From the depths of my skeptical and rationalist soul, I ask the Lord to protect me from California touchie-feeliedom. I must, nonetheless, express my concurrence with Siegel's important theme, though I hasten to express two vital caveats. First, I harbor no mystical notions about the potential value of mental calm and tenacity. We do not know the reasons, but I am confident that explanations will fall within the purview of scientific accessibility (and will probably center on how the biochemistry of thought and emotion feed back upon the immune system). Second, we must stand resolutely against an unintended cruelty of the "positive attitude" movement—insidious slippage into a rhetoric of blame for those who cannot overcome their personal despair and call up positivity from some internal depth. We build our personalities laboriously and through many years, and we cannot order fundamental changes just because we might value their utility: no button reading "positive attitude" protrudes from our hearts, and no finger can coerce positivity into immediate action by a single and painless pressing. How dare we blame someone for the long-standing constitution of their tendencies and temperament if, in an uninvited and unwelcome episode of life, another persona might have coped better? If a man dies of cancer in fear and despair, then cry for his pain and celebrate his life. The other man, who fought like hell and laughed to the end, but also died, may have had an easier time in his final months, but took his leave with no more humanity.

My own reaction to reading this chillingly pessimistic literature taught me something that I had suspected, but had not understood for certain about myself (for we cannot really know until circumstances compel an ultimate test): I do have a sanguine temperament and a positive attitude.

I confess that I did sit stunned for a few minutes, but my next reaction was a broad smile as understanding dawned: "Oh, so that's why she told me not to read any of the literature!" (My doctor later apologized, explaining that she had erred on the side of caution because she didn't yet know me. She said that if she had been able to gauge my reaction better, she would have photocopied all the reprints and brought them to my bedside the next day.)

My initial burst of positivity amounted to little more than an emotional gut reaction—and would have endured for only a short time, had I not been able to bolster the feeling with a genuine reason for optimism based upon better analysis of papers that seemed so brutally pessimistic. (If I had read deeply and concluded that I must inevitably die eight months hence, I doubt that any internal state could have conquered grief.) I was able to make such an analysis because my statistical training, and my knowledge of natural history, had taught me to treat variation as a basic reality, and to be wary of averages—which are, after all, abstract measures applicable to no single person, and often largely irrelevant to individual cases. In other words, the theme of this book—"full house," or the need to focus upon *variation within entire systems,* and not always upon abstract measures of average or central tendency—provided substantial solace in my time of greatest need. Let no one ever say that knowledge and learning are frivolous baubles of academic sterility, and that only feelings can serve us in times of personal stress.

I started to think about the data, and the crucial verdict of "eight months' median mortality" as soon as my brain started functioning again after the initial shock. And I followed my training as an evolutionary biologist. Just what does "eight months median mortality" signify? Here we encounter the philosophical error and dilemma that motivated this book. Most people view averages as basic reality and variation as a device for calculating a meaningful measure of central tendency. In this Platonic world, "eight months' median mortality" can only signify: "I will most probably be dead in eight months"—about the most chilling diagnosis anyone could ever read.

But we make a serious mistake if we view a measure of central tendency as the most likely outcome for any single individual—though most of us commit this error all the time. Central tendency is an abstraction, variation the reality. We must first ask what "median" mortality signifies. A median is the third major measure of central tendency. (I discussed the other two in the last chapter—the mean, or average obtained by adding all the values and dividing by the number of cases; and the mode, or most common value.) The median, as etymology proclaims, is the halfway point in a graded array of values. In any population, half the individuals will be below the median, and half above. If, say, in a group of five children, one has a penny, one a dime, one a quarter, one a dollar, and one ten dollars, then the kid with the quarter is the median, since two have more money and two less. (Note that means and medians are not equal in this case. The mean wealth of $2.27—the total cash of $11.36 divided by five—lies between the fourth and fifth kids, for the tycoon with ten bucks overbalances all the paupers.) We favor medians in such cases, when extension at one end of the variation drags the mean so far in that direction. For mortality in mesothelioma and other diseases, we generally favor the median as a measure of central tendency because we want to know the halfway point in a series of similar outcomes graded in time. A higher mean might seem misleading in the case of mesothelioma because one or two people living a long time (the analog of the kid with ten bucks) might drag the mean to the right and convey a false impression that most people with the disease will live for more than eight months—whereas the median correctly informs us that half the afflicted population dies within eight months of diagnosis.

We now come to the crux of practice: I am not a measure of central tendency, either mean or median. I am one single human being with mesothelioma, and I want a best assessment of *my own* chances—for I have personal decisions to make, and my business cannot be dictated by abstract averages. I need to place myself in the most probable region of the variation based upon particulars of my own case; I must not simply assume that my personal fate will correspond to some measure of central tendency.

I then had the key insight that proved so life-affirming at such a crucial moment. I started to think about the variation and reasoned that the distribution of deaths must be strongly "right skewed" in statistical parlance—that is, asymmetrically extended around a chosen measure of central tendency, with a much wider spread to the right than to the left. After all, there just isn't much room between the absolute minimum value of zero (dropping dead at the moment of diagnosis) and the median value of eight months. Half the variation must be scrunched up into this left half of the curve (see Figure 4) between the minimum and the median. But the right half may, in principle, extend out forever, or at least into extreme old age. (Statisticians refer to the ends of such distribution as "tails"—so I am saying that the left tail abuts a wall at zero survivorship, while the right tail has no necessary limit but the maximal human life span.)

I needed, above all, to know the form and expanse of variation, and my most probable position within the spread. I realized that all factors favored a potential location on the right tail—I was young, rarin' to fight the bastard, located in a city offering the best possible medical treatment, blessed with a supportive family, and lucky that my disease had been discovered relatively early in its course. I was therefore far more interested in the right tail (my probable residence) than in any measure of central tendency (an abstraction with no special relevance to my case). What, then, could possibly be more uplifting than an inference that the spread of variation would be strongly right skewed? I then checked the data and confirmed my supposition: the variation was markedly right skewed, with a few people living a long time. I saw no reason why I shouldn't be able to reside among these inhabitants of the right tail.

This insight gave me no guarantee of normal longevity, but at least I had obtained that most precious of all gifts at a crucial moment: the prospect of substantial time—to think, to plan, and to fight. I would not immediately have to follow Isaiah's injunction to King Hezekiah: "Set thine house in order: for thou shalt die, and not live." I had made a good statistical inference about the importance of variation and the limited utility of averages, and I had been able to confirm this suspicion with actual data. I had used knowledge and gained succor. (This story boasts an even more favorable outcome. I was destined for the right tail anyway, but an experimental treatment worked and has now probably removed the disease entirely. Old distributions offer no predictions for new situations. I trust that I am now headed for the right tail of a new distribution based on this successful treatment: death at a ripe old age in two high figures—maybe even three low ones.)

I present this tale not only for the pleasure of retelling a crucial yarn about my life, but because it encapsulates all the principles that form the

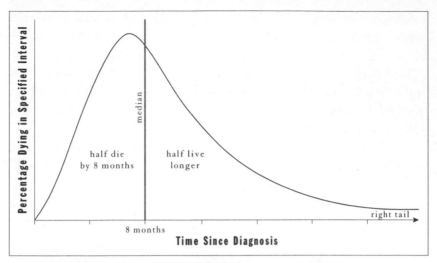

FIGURE 4

A right-skewed distribution for time of death for an illness with a median mortality of eight months. Each individual must be considered as a separate entity and the entire distribution cannot be characterized by its median value.

core of this book. First of all, my story illustrates the importance of *variation within whole systems* as an ultimate reality—and the limited utility (and abstract nature) of averages. Moreover, in a didactic sense for this book, my story embodies the three terms and concepts that form the conceptual apparatus for all the examples to follow. Let me try, then, to present these principles in a formal way, and in a context that will not seem too dry or forbidding.

THE SKEW OF A DISTRIBUTION. If we decide to treat variation as a principal reality, then we must consider the standard terms and pictures for portraying populations and their spread. We all know the conventional icon, called a frequency distribution, with the horizontal axis scaled as a graded series for the measure under consideration (height, weight, age, survivorship in disease, batting average, anatomical complexity, etc.), and the vertical axis scaled for the number of individuals in each interval of horizontal values (those weighing between ten and twenty pounds, between twenty and thirty, etc.; those between ten and fifteen years of age, between fifteen and twenty, etc.). Frequency distributions may be symmetrical— that is, with an identical shape and number on either side of the central tendency. The ubiquitous and idealized "normal distribution" or "bell curve" of current notoriety (Figure 5) is defined as symmetrical in this manner. We have all seen normal curves so often that we have been subtly led to treat natural systems as though they longed to display this ideal form. But most actual populations are not so simple or tidy. (Systems with purely random variation around a mean value will be symmetrical—as variation falls with equal probability on either side of the mean, with any single case more likely to lie close to the mean than far away. Runs of heads or tails in coin tossing, for example, form normal distributions. We regard the normal distribution as canonical because we tend to view systems as having idealized "correct" values, with random variation on either side—another consequence of lingering Platonism. But nature does not match our expectations very often.)

Actual distributions are often asymmetrical, or skewed. In a skewed distribution, as illustrated by my personal story, variation stretches out farther on one side than the other—called either "right" or "left" skewed de-

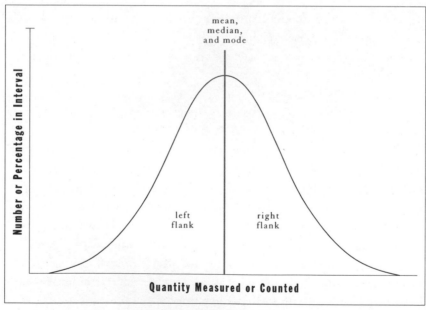

FIGURE 5

An idealized bell curve or normal frequency distribution, showing that all measures of central tendency (mean, median, and mode) coincide.

pending on the direction of elongation (Figure 6). The reasons for skewing are often fascinating and full of insight about the nature of systems—for skewing measures departure from randomness. Since this book treats the nature of variation, and the reasons for changes in spread through time, skewing becomes an important principle in all my examples.

MEASURES OF CENTRAL TENDENCY AND THEIR MEANING. I have discussed the three standard measures of central tendency, or "average" value—the mean (or conventional average calculated by adding all values and dividing by the number of cases), the median (or halfway point), and the mode (or most common value). In symmetrical distributions, all three measures coincide—for the center is, simultaneously, the most common

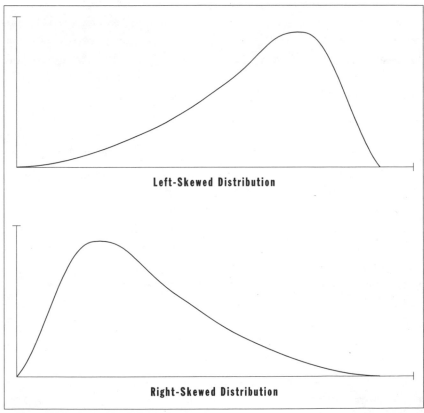

Left-Skewed Distribution

Right-Skewed Distribution

FIGURE 6
Left- and right-skewed distributions.

53

value, the halfway point (with equal numbers of cases on either side), and the mean. This coincidence, I suspect, has led most of us to ignore the vital differences among these measures, for we view "normal curves" as, well, normal—and regard skewed distributions (if we grasp the principle at all) as peculiar and probably rare. But measures of central tendency differ in skewed distributions—and a major source of employment for economic and political "spin doctors" lies in knowing which measure to choose as the best propaganda for the honchos who hired your gun.

I have already shown how the higher mean and lower mode of a right-skewed distribution in incomes can be so exploited (see page 37). In general, when a distribution is prominently skewed, mean values will be pulled most strongly in the direction of skew, medians less, and modes not at all. Thus, in right-skewed distributions, means generally have higher values than medians, and medians higher than modes. Figure 7 should make these relationships clear. If we start with a symmetrical distribution (with equal mean, median, and mode), and then pull the variation to form a right-skewed distribution, the mean will change most in the direction of skew—for one new millionaire on the right tail can balance hundreds of indigent people on the left tail. The median changes less, for a single pauper will now compensate the millionaire when we are only counting noses on either side of a central tendency. (The median might not move at all if only the wealth, and not the number, of people increases on the right side of the distribution. But if the number of wealthy people at the right tail increases as well, then the median will also shift to the right—but not so far as the mean.) The mode, meanwhile, may well stay put and not vary at all, as mean and median grow in an increasingly right-skewed distribution. Twenty thousand per year may remain the most common income, even while the number of wealthy people constantly increases.

"WALLS," OR LIMITS TO THE SPREAD OF VARIATION. As a major reason for skew, variation is often limited in the extent of potential spread in one direction (but much freer to expand in the other). The reasons for such limits may be trivial or logical—as in my cancer story where a person can't die of mesothelioma before he gets the disease, and zero time between onset and death therefore becomes an irreducible minimum. The reasons may also be subtle and more interesting—as in the examples of batting averages and life's history to be presented in Parts Three and Four of this book.

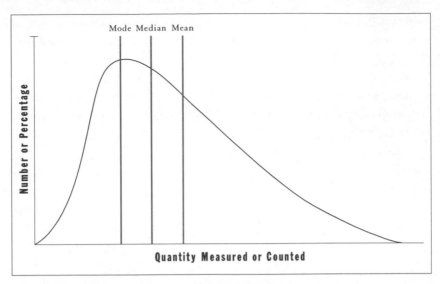

FIGURE 7

In a right-skewed distribution, measures of central tendency do not coincide. The median lies to the right of the mode, and the mean lies to the right of both other measures.

In either case, such limits often produce skewed distributions, because variation can expand in only one direction—you can't die of mesothelioma before you get it, but you can live for years and years after a diagnosis. With an eight-month median mortality, and a rigid lower limit at time zero, how could the distribution of deaths be anything but strongly right skewed?

Throughout this book, I shall refer to such limits upon the spread of variation as "walls"—either "right walls" or "left walls" depending upon their position. Left walls induce right-skewed distributions (for variation is only free to expand away from a wall); right walls provoke left-skewed distributions. The left wall of my cancer story leads to a right-skewed distribution of deaths.

(I have considered the cultural bias involved in the largely arbitrary designation of right as the direction for higher values, left for lower—though, depending upon the example, lower may be judged better, as in distributions for weight in our diet-conscious society. I suppose that we fall into this bias for two reasons, one insidious and the other benign. Prejudice against our left-handed minority—an old and probably universal feature of human cultures, I fear—must set the major reason. Jesus sits *ad dextram patris,* at the right hand of the father. Right, etymologically, is dextrous—and "law" is *droit* in French and *Recht* in German, both meaning right. Left is both sinister and gauche. For the benign reason, we read from left to right and therefore conceptualize growth and increase in this direction. Were I writing this book in Israel, which also has a right-handed majority, I might think of left walls as directions of increase. Were I writing in Japan, I might be talking of top and bottom walls. So be it.)

Readers need to grasp only these three nontaxing concepts about the nature of variation in order to render all the examples of this book fully digestible—right and left walls as limits to the spread of variation; right- and left-skewed distributions arising as results of these limits; and differences among means, medians, and modes as measures of central tendency.

·5·

Case Two: Life's Little Joke

Genuine changes in central tendency are meaningful, but our failure to consider variation has led to a backwards interpretation: the evolution of horses

The most erroneous stories are those we think we know best—and therefore never scrutinize or question. Ask anyone to name the most familiar of all evolutionary series and you will almost surely receive, as an answer: horses, of course. The phyletic racecourse from small, many-toed protohorses with the charming name eohippus, to a big, single-toed Clydesdale hauling the Budweiser truck, or Man O' War thundering down the stretch, must be the most pervasive of all evolutionary icons. Does any major museum not have a linear series of cases against a long wall, or up the center of a main hall, one skeleton in each, and all illustrating the triumphant trend?

This horse story also represents the oldest of established evolutionary series—a major reason for its fame. Thomas Henry Huxley himself, Darwin's most celebrated supporter, first proposed the sequence from European fossils in 1870. This original version did not long survive because Huxley's three European fossils, linked as an evolutionary series, actually represent three separate migrations of American stocks, with extinction in Europe following each incursion. Meanwhile, the full story was unfolding in America.

In 1876, Huxley made his only voyage to the United States, primarily to participate in celebrations for our Centennial and, in particular, to give the principal address at the founding of Johns Hopkins University. He visited Othniel C. Marsh, America's leading vertebrate paleontologist, to see the magnificent series of fossil horses that Marsh had gathered in the American West. Marsh convinced him that the American series formed a true evolutionary main line, with the European offshoots as disconnected side branches. Huxley had to scramble, for he had promised to give a lecture in New York, just a few weeks later, on fossil horses—and he now had to revise his story completely.

Marsh agreed to help with these quick changes, and he prepared a famous chart for Huxley's use in the New York lecture (reproduced here as Figure 8). This figure, among the most celebrated in the history of science, shows two of the three major trends in our classic tale: (1) reduction in number of toes, from four on the front feet and three behind in the earliest horses (bottom of the figure), to three functional toes, to a central toe with two shortened side toes, to a single toe with two side splints as vestiges of former toes (modern horses at top of the figure); (2) steady increase in the height of molar teeth (fifth column of specimens) with elaboration in the convolutions of their cusps (columns six and seven). Marsh chose to draw all his specimens at the same size, and therefore didn't show the third and most evident trend of marked increase in bulk from the initial stage (which he described as cat-sized, though fox terriers have since triumphed as a canonical metaphor—see Gould, 1991) to the massive Clydesdale of today. Later versions showed all three coordinated trends, as in the best-known figure by the next generation's paleontological leader, William D.

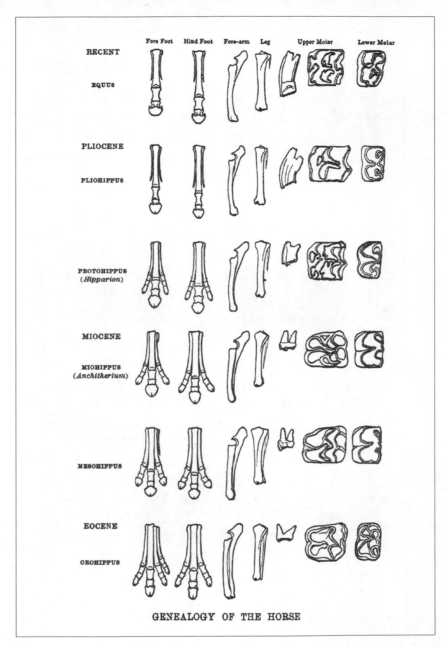

Fore Foot   Hind Foot   Fore-arm   Leg   Upper Molar   Lower Molar

RECENT

EQUUS

PLIOCENE

PLIOHIPPUS

PROTOHIPPUS
(*Hipparion*)

MIOCENE

MIOHIPPUS
(*Anchitherium*)

MESOHIPPUS

EOCENE

OROHIPPUS

GENEALOGY OF THE HORSE

FIGURE 8
A famous chart on the evolution of horses prepared by O. C. Marsh for T. H. Huxley's
New York lecture. Note the linear march to progress in all characters.

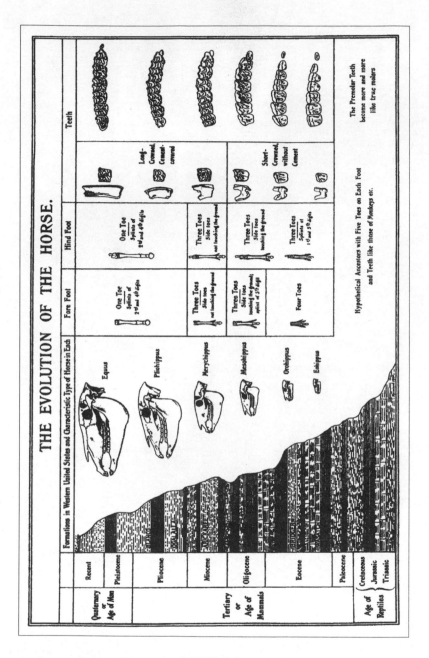

FIGURE 9

W. D. Matthew's linear and progressive evolution of horses plotted in stratigraphic order, showing increasing body size, decreasing number of toes, and increasing height of teeth.

Matthew, first published early in our century in a pamphlet by the American Museum of Natural History, still for sale in the museum shop during my youth in the 1950s, and endlessly reproduced all along. (One copy, for example, appeared in the textbook used by John Scopes to teach evolution to the schoolchildren of Dayton, Tennessee—a source, therefore, for W. J. Bryan's fulminations at Scopes's famous "monkey" trial—"no more repulsive doctrine was ever proclaimed by man.") This version arranges the specimens in stratigraphic order next to a geological column and shows all trends of size, toes, and teeth (Figure 9).

In some legitimate though limited sense, these trends are true. The first horses, technically called *Hyracotherium* (though I love the informal, if taxonomically incorrect, name eohippus, or "dawn horse"), were small, and did have four toes in front, three behind, and low-crowned teeth. The standard story for the advantages of these trends—probably also basically correct—points to a switch in habitat from browsing in forested areas (where many toes hug the soft ground and low-crowned teeth can manage the leafy vegetation) to grazing on plains (where hooves are superior on the hard terrain, and strong, high-crowned teeth deal better with tough grasses and their substantial content of silica. Grasses first evolved in the midst of equine evolution, thus promoting these trends by opening up an extensive new habitat.) In a strictly join-the-dots sense, we do make a correct statement about genealogy when we connect the point for *Hyracotherium* with the point for modern *Equus* (the only living genus of horses, including eight species—three zebras, four donkeys and asses, and Old Dobbin, or *Equus caballus,* representing true horses alone).

So far, so good—but (as I shall show) so very limited, and so misleading. The lineage of *Hyracotherium* to *Equus* represents only one pathway through a very elaborate bush of evolution that waxed and waned in a remarkably complex pattern through the last 55 million years. This particular pathway cannot be interpreted as a summary of the bush; or as an epitome of the larger story; or, in any legitimate sense, as a central tendency in equine evolution. We have chosen this little sample of a totality for one reason alone: *Equus* is the only living genus of horses, and therefore the only *modern* animal that can serve as an endpoint for a series. If you are committed to depicting the evolution of any living group as a single pathway from an ancestral point to an item of current glory, then I suppose that the story must be told in this conventional way. But when we consider more comprehensive models of evolution, we must call such pictures into question.

We therefore arrive at my favorite subject of ladders versus bushes, or, in the context of this book, individual pathways chosen with prejudice versus entire systems (full houses) and their complete variation. As the Bible says about wisdom, so too may we state about the proper iconography of evolution: "She is a tree of life to them that lay hold upon her." Evo-

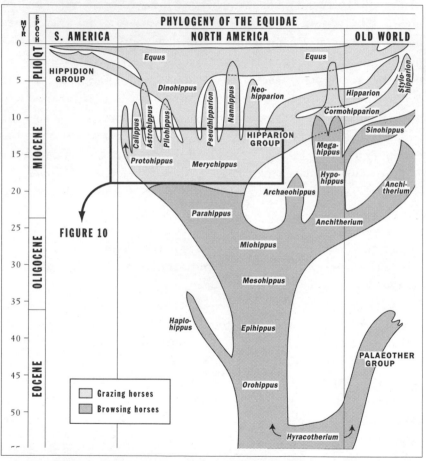

FIGURE 10

The more complex branching evolution of horses as depicted by Bruce MacFadden in 1988.

lution rarely proceeds by the transformation of a single population from one stage to the next. Such an evolutionary style, technically called *anagenesis,* would permit a ladder, a chain, or some similar metaphor of linearity to serve as a proper icon of change. Instead, evolution proceeds by an elaborate and complex series of branching events, or episodes of speciation (technically called *cladogenesis,* or branch-making). A trend is not a march along a path, but a complex series of transfers, or side steps, from one event of speciation to another. The evolutionary bush of horses includes many terminal tips, and each leads back to *Hyracotherium* through a labyrinth of branching events. No route to *Hyracotherium* is straight, and none of the numerous labyrinthine paths has any special claim to centrality (see Figure 10). We run a steamroller right over a fascinatingly complex terrain when we follow the iconographic convention for displaying the pathway from *Hyracotherium* to *Equus* as a straight line.

So why do we engage in such distortion, and why have horses become the standard example of an evolutionary "trend"? At this point in the argument, we encounter the irony that I have called "life's *little* joke" (see Gould, 1987). We choose horses because their living species represent the endpoint of such an *unsuccessful* lineage. The situation is even "worse," and fully subject to generalization: our bias against considering the variation of full systems, and trying instead to depict trends as "entities moving somewhere," virtually guarantees that all our standard examples of evolutionary movement and "progress" must feature failing groups, so reduced from earlier bushiness that only a single twig—life's *little* joke—survives as a relic of former glory.

What are the real success stories of mammalian evolution? We can answer this question without ambiguity, at least in terms of numerous species and vigorous radiation: rats, bats, and antelopes (or, in more formal terms, the orders Rodentia, Chiroptera, and the family Bovidae among the artiodactyls). These three groups dominate the world of mammals, both in numbers and in ecological spread. Yet has anyone ever seen an iconographic depiction of their success?

We never feature these groups because we do not know how to draw their triumph. Evolution, to us, is a linear series of creatures getting bigger, fancier, or at least better adapted to local environments. When groups are truly successful, and their tree contains numerous branches, all prospering at once, we can designate no preferred pathway—and we there-fore have no convention for depicting, or even (really) for conceiving, their evolution. But when an evolutionary bush has been so pruned by extinction that only one lineage survives—a twig from an earlier arborescence, a sliver of former copiousness—then we can fool ourselves into viewing this tiny remnant as a unique culmination. We either forget that other pathways to extinct lineages once existed, or we scorn them as "dead ends"—irrelevant side branches from a supposed main trunk. We then bring forth our conceptual steamroller to straighten out the little path from the surviving twig to the ancestral stock—and, finally, with the positive spin of a consummate evolutionary trendmaker, we praise the progress of horses.

Many classic "trends" of evolution are stories of such unsuccessful groups—trees pruned to single twigs, then falsely viewed as culminations rather than lingering vestiges of former robustness. We cannot grasp the irony of life's little joke until we recognize the primacy of variation within complete systems, and the derivative nature of abstractions or exemplars chosen to represent this varied totality. The full evolutionary bush of horses is a complete system; the steamrollered "line" from *Hyracotherium* to *Equus* is one labyrinthine path among many, with no special claim beyond the fortuity of a barely continued existence.

These conceptual errors have plagued the interpretation of horses, and the more general evolutionary messages supposedly conveyed by them, from the very beginning. Huxley himself, in the printed version of his capitulation to Marsh's interpretation of horses as an American tale, used the supposed ladder of horses as a formal model for all vertebrates. For example, he denigrated the teleosts (modern bony fishes) as dead ends without issue (1880, page 661): "They appear to me to be off the main line of evolution—to represent, as it were, side tracks starting from certain points of that line." But teleosts are the most successful of all vertebrate groups. Nearly 50 percent of all vertebrate species are teleosts. They stock the world's oceans, lakes, and rivers, and include nearly one hundred times as many species as primates (and about five times more than all mammals combined). How can we call them "off the main line" just because we can trace our own pathway back to common ancestry with theirs, more than 300 million years ago?

W. D. Matthew, author of the most famous icon for the equine ladder (Figure 9), fell into the same error because his designation of one pathway as a main line forced him to interpret all others as diversions of lesser value. Matthew (1926, page 164) called his ladder "the direct line of succession," and added that "there are also a number of side branches, more or less closely related." But Matthew then imposed a brand of near indecency upon his previous charge of mere laterality, as he described (1926, page 167) "a number of side branches leading up . . . to aberrant specialized Equidae now extinct." But in what way are these extinct lineages more specialized than a modern horse, or in any sense more peculiar? Their phyletic death sets the only possible rationale for a designation of aberrancy, but more than 99 percent of all species that ever lived are now extinct—and disappearance is not the biological equivalent of a scarlet letter.

I have thus far presented the case of horses only as a general argument about bushiness versus linearity. I do not deny the factuality of the conventional pathway and its trends in size, teeth, and toes, but I do wish to demonstrate what a distorted—even backward—view this little piece provides when we depict *Hyracotherium* → *Equus* as the essence of the history of horses, and then ignore the variation supplied by a myriad of other pathways in the full house of the equine bush. Three categories of detail should cement the importance of the opposite perspective gained by considering the changing spread of variation through time—horses as a declining lineage within a failing larger group.

1.   The evolutionary tree of horses is copiously bushy throughout; no geological segment of equine history can be construed as featuring a wide main line with dribbling side branches. Bruce MacFadden of the Florida Museum of Natural History, our leading modern expert on the paleobiology of horses, recently published a simplified picture of the equine tree (reproduced here as Figure 10). Consider the last 20 million years from the key transition of browsing to grazing, up to modern times. We note only copious branching bushiness, and nothing that anyone could identify as a central thrust amid the diversity. MacFadden couldn't even begin to depict all the complex branching in his single diagram, so he expanded a key portion (as indicated by the square that he drew over Figure 10), and presented a fuller account of this 7-million-year interval (reproduced here as

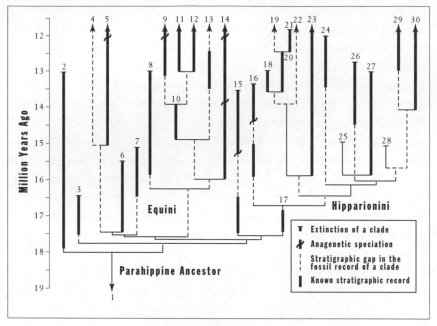

FIGURE 11

So much branching occurs in the evolution of horses during the Middle Miocene that Mac-Fadden, as shown in Figure 10, could not include all the individual lineages. This figure is an enlargement of the box shown in Figure 10. Note how many branching events occurred during this relatively short interval.

Figure 11). In North America alone, between 15 and 18 million years ago, at least nineteen species originated by branching. By 15 million years ago, sixteen contemporaneous grazing species inhabited North America (while several older lineages of browsing horses also lived in America and the Old World). This diversity hardly changed during the next 7 million years "as extinctions balanced originations, resulting in a steady-state diversity pattern" (MacFadden, 1988, page 2). North American diversity then declined rapidly, and the entire bush of horses eventually died out in the New World. (Remember how Cortez's horses terrified the Aztecs, who had never seen the beasts that had originated on their continent and then become extinct. Eurasia was an outpost of equine survival, not a center for an expanding trend.)

Two points stand out in this epitome for the last third of equine history. First we note a primary signal of branching, branching, and more branching. Where, in this forest, could anyone identify a main trunk? The bush has many tips, though all but one, the genus *Equus,* are extinct. Each tip can be connected to a last common ancestor by a labyrinthine route, but no paths are straight, and all lead back by sidestepping from one event of branching speciation to another, and not by descent down a ladder of continuous change. If you venture an argument that the pathway to modern *Equus* should be viewed as a main line because the genus still lives and once spread (by its own devices, and not by human transport) over all major continents, I reply that *Equus* died out over most of its range, including the North American fatherland, and that all modern species derive from Old World remnants. Second, I think that any unbiased observer must identify decline as the major feature of equine evolution during the last 10 million years—the very period when traditional ladder models proclaim perfection and fine-tuning of the distinctive trend to a single hoofed toe, with side toes reduced to vestigial splints. An average of sixteen contemporaneous species lived in North America alone from about 15 million to about 8 million years ago—until, to invoke Agatha Christie's famous image, they died one by one—and then there were none.

Rearguard defenders of the ladder might reply that I have been discussing only the last (and admittedly bushy) third of equine evolution. What about the first 40 million years, shown as tolerably linear even on MacFadden's arborescent picture (Figure 10)? This earlier period has been the chief domain for friends of linearity. Even G. G. Simpson, who began the transition to bushy thinking in his wonderful 1951 book, *Horses,* and who drew the first famous arborescent diagram of equine phylogeny (a less bushy ancestor of MacFadden's version, reproduced here), defended the basic linearity of this earlier record. "The line from *Eohippus* [*Hyracotherium*] to *Hypohippus,"* he wrote (1951, page 215), "exemplifies a fairly continuous phyletic evolution." Simpson especially emphasized the supposedly gradual and continuous transformation from *Mesohippus* to *Miohippus* near the top of this sequence (see Figure 10 for all names and times):

> The more progressive horses of the middle Oligocene . . .
> are placed by convention in a separate genus, *Miohippus.*
> In fact *Mesohippus* and *Miohippus* intergrade so perfectly
> and the differences between them are so slight and vari-

able that even experts find it difficult, at times nearly impossible, to distinguish them clearly.

The enormous increase in fossil evidence since Simpson's time has allowed paleontologists Don Prothero and Neil Shubin (1989) to falsify this view, and to introduce extensive bushiness into this last stronghold of the ladder, as predicted by the theory of punctuated equilibrium (see Eldredge and Gould, 1972; Gould and Eldredge, 1993). Prothero and Shubin made four major discoveries in this early segment of equine history that Simpson had designated as the strongest case for a gradual sequence of linear transformation—the transition from Mesohippus to Miohippus.

First, the two genera can be sharply distinguished by features of the footbones, previously undiscovered. *Mesohippus* does not grade insensibly into *Miohippus*. (Previous claims had been based on teeth, the best preserved parts of mammalian skeletons. The genera cannot be distinguished on dental evidence—the major criterion available to Simpson.)

Second, *Mesohippus* does not evolve to *Miohippus* by insensible degrees of gradual transition. Rather, *Miohippus* arises by branching from a *Mesohippus* stock that continues to survive long afterward. The two genera overlap in time by at least 4 million years.

Third, each genus is itself a bush of several related species, not a rung on a ladder. These species often lived and interacted in the same area at the same time. One set of strata in Wyoming, for example, has yielded three species of *Mesohippus* and two of *Miohippus,* all contemporaries.

Fourth, the species of these bushes tend to arise with geological suddenness, and then to persist with little change for long periods. Evolutionary change occurs at the branch points themselves, and trends are not continuous marches up ladders, but concatenations of increments achieved at nodes of branching on evolutionary bushes. Prothero and Shubin write,

> This is contrary to the widely held myth about horse species as gradualistically varying parts of a continuum, with no real distinctions between species. Throughout the history of horses, the species are well-marked and static over millions of years. At high resolution, the gradualis-

tic picture of horse evolution becomes a complex bush of
overlapping, closely related species.

In other words, bushiness now pervades the entire phylogeny of horses.

2.   Plausible alternative histories would have yielded a very different and
not nearly so attractive story. The substitution of bushes for ladders certainly calls into question, but does not necessarily falsify, the conventional
lockstep view of transitions to fewer toes, larger bodies, and higher-
crowned teeth. After all, older branches of a bush need not endure for long,
and their early removal would leave no ancient vestiges to compromise a
trend by persistent variation. If all the early branches die, and all the later
twigs bear "progressive" features, then the tree becomes "modernized"
throughout—and we may fairly talk of a pervasive trend. If all small
horses die early, if no three-toed horses survive into the regime of one-toed
Equus, then we may justly speak of general trends to increased size and a
single hoof—and the old marching order from *Hyracotherium* to *Equus*
might be defended as a fair epitome of real directionality (while still subject to criticism for neglecting the equally important pattern of waxing and
waning diversity). In such a world, the objections that I have raised would
be carping and trivial. Yes, we could still emphasize that many pathways
run through the bush, and that *Hyracotherium* to *Equus* marks only one
lineage—but if all pathways pass through the same sequence to larger size
and fewer toes, then any one will show the genuine trend, and we shouldn't
be too critical if convention favors one case over all others.

      This last-ditch defense of equine progress cannot be sustained. The
conventional trends are by no means pervasive (though their relative frequency does increase through the bush, albeit in a fitful way). Several late
lineages negate the most prominent trends, and a different outcome for
the history of horses—perfectly plausible in our world of contingency (see
Gould, 1989)—would have compelled a radically altered tale.

      Consider just one arresting scenario. Contrary to the usual view that
horses increase inexorably in body size, MacFadden (1988) studied all
ancestral-descendant pairs of species that he could identify with confidence
on the equine bush. Of twenty-four such pairs, he found that five, or more

than 20 percent, showed a *decrease* in size. Dwarfing has been a common and persistent phenomenon, repeated throughout the history of horses. Even the first genus, *Hyracotherium,* included periods of size decrease during its geological history (see Gingerich, 1981).

The most recent, and most profound, trend to dwarfing occurred in a North American genus appropriately named *Nannippus* (or dwarfed horse). Simpson writes of this remarkable genus (1951, page 140): "Some of the late specimens were miniatures no higher than a small Shetland pony and considerably more slender. These graceful creatures had long, thin legs and feet, and the general form probably suggested a small gazelle more than an ordinary horse."

Now suppose that *Nannippus* had survived as the only living member of the Equidae, and *Equus* had died or never arisen. How would we then tell the story of horses in our biased mode of running steamrollers over one pathway through the bush and calling the resulting line canonical? I hear you crying "foul." You say that *Nannippus* was a funny little side branch and *Equus* a powerful main line—so I must be playing verbal games with a story that could never have occurred. Not so; my tale is plausible, but just unrealized. *Nannippus* showed substantial geographic breadth and geological depth. The genus lived in the United States and Central America, arose more than 10 million years ago, and failed to survive by only a whisker, becoming extinct only about 2 million years ago (MacFadden and Waldrop, 1980). Four species have been described (MacFadden, 1984), and their range of some 8 million years greatly exceeds the longevity of *Equus* (see Figure 11). If you say that *Equus* had a greater chance because this modern genus spread from an American homeland into Eurasia and Africa, while *Nannippus* never colonized the Old World, I reply that *Equus* became entirely extinct throughout its hemisphere of origin, and therefore only survived by a whisker itself. Suppose that *Nannippus* had migrated and *Equus* stayed at home?

What would be left of our vaunted horse story if *Nannippus* had survived, and *Equus* died? We wouldn't be advertising any drive to increased size because *Nannippus,* though a dwarfed descendant of larger ancestors, isn't much bigger than the original *Hyracotherium.* We wouldn't be getting very excited about reduction in toes either, because *Nannippus* still sported three on each foot (though the side toes were reduced), whereas

the original *Hyracotherium* had four toes on the front feet and three behind (not five on each limb, as commonly misconstrued). We would be left,
in fact, only with the trend to increased crown height of the molar teeth—
and here we could gloat, because *Nannippus* chewed with the relatively
tallest teeth of any horse in history, including modern *Equus*. But then,
tooth height has never provided much of a draw for museums or textbook
diagrams, and the conventional story rests upon reduction of toes and
growth of body. In short, if *Nannippus* had survived and *Equus* died, we
wouldn't be telling any famous story about horses at all. The equine bush
would become just another anonymous part of the rich mammalian record,
known to specialists and unadvertised to the public. Yet nothing would
be different but the substitution of one twig for another at the very end of
a rich history.

3.   Modern horses are not only depleted relative to horses of the past; on
a larger scale, all major lineages of the Perissodactyla (the larger mammalian group that includes horses) are pitiful remnants of former copious
success. Modern horses, in other words, are failures within a failure—
about the worst possible exemplars of evolutionary progress, whatever
such a term might mean.

   Mammals are ranked into some twenty major divisions, called orders.
Horses belong to the order Perissodactyla, or odd-toed ungulates—large,
herbivorous animals with an odd number of toes on each foot. (The other
major ungulate order, called Artiodactyla, contains creatures with an even
number of toes on each foot. Each of these orders represents a genuine evolutionary unit traceable to a common ancestor, not an artificial construct
devised only by counting toes.) The perissodactyls are a small and depleted order, with only three surviving groups, and seventeen species *in
toto*—horses (eight species), rhinoceroses (five species), and tapirs (four
species).

   If you become overly sanguine and insist that you won't demerit this
group for limited modern diversity because the three kinds of survivors
fascinate us so much, I can only recommend a deeper geological look and
the famous lamentation of David over Saul and Jonathan: "How are the
mighty fallen in the midst of the battle." Perissodactyls were once the giants of mammalian life; we now honor a few straggling ghosts in our zoos

because they intrigue us, and because one species has made such a profound difference in human history.

The rhinocerotoids were once among the most abundant and varied of all mammalian groups. Their extensive ecological range included small and sleek running forms no larger than a dog (the hyracodontines), rotund river-dwellers that looked like hippos (teleoceratines), an array of dwarfed forms, and the largest land mammals that ever lived—the giant indricotheres, including the all-time size champion *Paraceratherium* (often called *Baluchitherium*), which stood eighteen feet tall at the shoulder and browsed on treetops (see Prothero, Manning, and Hanson, 1986; Prothero and Schoch, 1989; Prothero, Guérin, and Manning, 1989). The five modern species, all looking much alike, all Old World, and all endangered, form a sad remnant of former glory. The same story may be told for horses, with their decline from sixteen to zero Old World species; and for tapirs, with their modern Asian and South American remnants of a former worldwide spread.

Moreover, the three living lineages include only a fraction of former perissodactyl diversity, for several major groups have been lost entirely— including, most spectacularly, the large-bodied and prominently horned titanotheres of early Tertiary times, and the chalicotheres, with their powerful digging claws.

Steady perissodactyl decline has been matched by a reciprocal rise to dominance of the contrasting artiodactyls, once a small group in the shadow of ruling perissodactyls, and now the most abundant order, by far, of large-bodied mammals. The perissodactyls survive as three twiggy vestiges. Artiodactyls are the lords of largeness—cattle, sheep and goats, deer, antelopes, pigs, camels, giraffes, and hippos. Need any more be said? Horses are remnants of a remnant, yet their story provides our false icon of progress—life's little joke. Antelopes represent the most vigorous family in an expanding and dominant group—but who has ever seen a picture of this group's astonishing success? Antelopes are examples of nothing in our museums and textbooks.

I therefore submit that the history of any entity (a group, an institution, an evolutionary lineage) must be tracked by changes in the variation of all components—the full house of their entirety—and not falsely epitomized as a single item (either an abstraction like a mean value, or a supposedly typical example) moving on a linear pathway. As a final footnote to life's little joke, I remind readers that one other prominent (or at least parochially beloved) mammalian lineage has an equally long and extensive history of conventional depiction as a ladder of progress—yet also lives today as the single surviving species of a formerly more copious bush. Look in the mirror, and don't be tempted to equate transient domination with either intrinsic superiority or prospects for extended survival.

Part Three

...

THE MODEL BATTER: EXTINCTION OF 0.400 HITTING AND THE IMPROVEMENT OF BASEBALL

·6·

Stating the Problem

During my lifetime, two events clearly stand out above all others as milestones in the history of batting in baseball: Joe DiMaggio's fifty-six-game hitting streak (see page 32), and Ted Williams's seasonal batting average of 0.406. Unfortunately, I missed them both because I was too busy gestating during the season of their joint occurrence in 1941. Boston Red Sox manager Joe McCarthy had offered to let Williams sit out the meaningless doubleheader of the season's last day (the Yankees had clinched the pennant long before). Williams's average stood at 0.3995, and would have rounded up to an even 0.400. No one had hit 0.400 for ten years, since New York Giants first baseman Bill Terry reached 0.401 in 1930. Ted couldn't bear to back in. He played both games, went 6 for 8, and finished the season at 0.406. No one has hit 0.400 since then (closest calls include George Brett at 0.390 in 1980, Rod Carew at 0.388 in 1977, and Ted Williams himself at 0.388, sixteen years later in 1957, the season of his thirty-ninth birthday). So I'm still waiting to see for myself what life *in utero* denied to my conscious understanding—and I'm not getting any younger.

Between 1901, when the American League began and Nap Lajoie hit 0.422, and 1930, when Terry hit 0.401, batting 0.400, while always honored, cannot be called particularly rare. League-leading averages exceeded 0.400 in nine of these thirty years, and seven players (Nap Lajoie, Ty Cobb, Shoeless Joe Jackson, George Sisler, Rogers Hornsby, Harry Heilmann, and Bill Terry) reached this apogee, three times each for Cobb and Hornsby. (Hornsby's 0.424 in 1924 tops the charts, while three players exceeded 0.400 in 1922—Sisler and Hornsby in the National League, Cobb in the American. I am, by the way, omitting the even more common nineteenth-century averages in excess of 0.400, because differing rules and practices in baseball's professional infancy make comparison difficult.) Then the bounty dried up: the thirties were a wasteland (despite high league averages during this decade, as I shall show later); Williams reached his lonely pinnacle in 1941; since then, zip.

If philately attracts perforation counters, and sumo wrestling favors the weighty, then baseball is the great magnet for statistical mavens and trivia hounds. Consider baseball's virtues for the numerically minded: Where else can you find a system that has operated with unchanged rules for a century (thus permitting meaningful comparison throughout), and has kept a complete record of all actions and achievements subject to numeration? Moreover, unlike almost any other team sport, baseball's figures are records of individual achievements, not elusive numbers that may be assigned to a single player, but really record some aspect of team play—for baseball is a congeries of contests between two individuals: hitter versus pitcher, runner versus fielder. Thus, records assigned to players of the past can be read as their personal achievement, and can be compared directly with the same measures of modern performers. No wonder, then, that the largest organization of scholarly fans, the Society for American Baseball Research, is so numerically minded, and has contributed, through its acronym, a new word to our language: sabermetrics, for the statistical study of sporting records.

Humans, as I have argued, are trend-seeking creatures (perhaps I should say "storytelling animals," for what we really love is a good tale—and, for reasons both cultural and intrinsic, we view trends as stories of the best sort). We are therefore driven to scan the charts of baseball records for apparent trends—and then to devise stories for their causes. Remember that our cultural legends include two canonical modes for trending: advances to something better as reasons for celebration, and declines to an abyss as sources of lamentation (and hankering after a mythical golden age of "good old days"). Since 0.400 hitting is both so noticeable and so justly celebrated, and since its pattern of decline and disappearance so clearly embodies the second of our canonical legends, no other trend in baseball's statistical history has attracted such notoriety and engendered such lamentation.

The problem seems so obvious in outline: something terrific, the apogee of batting performance, was once reasonably common and has now disappeared. Therefore, something profoundly negative has happened to hitting in baseball. I mean, how else could you possibly read the evidence? The best is gone, and therefore something has gotten worse. I devote this chapter to the paradoxical claim that extinction of 0.400 hitting really measures the general improvement of play in professional baseball. Such a claim cannot even be conceived while we remain stuck in our usual Platonic mode of viewing 0.400 hitting as a "thing" or "entity" in itself—for the extinction of good items must mean that something has turned sour. I must therefore convince you that this basic conceptualization is erroneous, and that you should not view 0.400 hitting as a thing at all, but rather as the right tail in a full house of variation.

· 7 ·

Conventional Explanations

More ink has been spilled on the disappearance of 0.400 hitting than on any other statistical trend in baseball's history. The particular explanations have been as varied as their authors, but all agree on one underlying proposition: that the extinction of 0.400 hitting measures the worsening of something in baseball, and that the problem will therefore be solved when we determine what has gone wrong.

This chorus of woe may be divided into two subchoirs, the first singing a foolish tune that need not long detain us, the second more worthy of our respect as an interesting error reflecting the deeper mistake that made this book necessary. The first explanation invokes the usual mythology about good old days versus modern mollycoddling, Nintendo, power lines, high taxes, rampant vegetarianism, or whatever contemporary ill you favor for explaining the morally wretched state of our current lives. In the good old days, when men were men, chewed tobacco, and tormented homosexuals with no fear of rebuke, players were tough and fully concentrated. They did nothing but think baseball, play baseball, and live baseball. Just look at Ty Cobb, sliding into third, spikes high (and directed at the fielder's flesh). How could any modern player, with his high salary and interminable distractions, possibly match this lost devotion? I call this version the Genesis Myth to honor the appropriate biblical passage about wondrous early times: "There were giants in the earth in those days" (Genesis 6:4). I don't think that we need to take such fulminations seriously (I shall give my reasons a bit later). For salaries in millions that can last for only a few years of physical prime and be lost forever in a careless moment, modern players can muster quite ferocious dedication to their craft; modern ballplayers certainly take better care of their bodies than any predecessor ever contemplated in the good old days of drinking, chewing, and womanizing.

The second, and more serious, approach tries to identify a constellation of factors that has made batting more difficult in modern times, and therefore caused the drop-off in leading averages. I shall argue that, while several of these explanations correctly identify new impediments to hitting, the premise of the entire argument—that disappearance of 0.400 hitting can only be tracking the decline of batting skills (either absolute or only relative)—is flat wrong. The extinction of 0.400 hitting measures the general improvement of play.

The Genesis Myth finds greatest support, unsurprisingly, among the best hitters of a more disciplined (and less remunerative) age who must suffer the self-aggrandizing antics of their modern, but now multimillionaire, counterparts. Ted Williams, the last 0.400 hitter, told reporters why his feat will not be soon repeated (*USA Today,* February 21, 1992): "Modern players are stronger, bigger, faster and their bodies are a little better than those of thirty years ago. But there is one thing I'm sure of and that is the average hitter of today doesn't know the little game of the pitcher and the hitter that you have to play. I don't think today there are as many smart hitters."

In his 1986 book, *The Science of Hitting,* Williams made the same claim, and explicitly embraced the key postulate of the Genesis Myth by stating that, since baseball hadn't altered in any other way, the decline of high hitting must record an absolute deterioration of batting skills among the best:

> After four years of managing . . . the one big impression
> I got was that the game hadn't changed. . . . It's basically
> the same as it was when I played. I see the same type
> pitchers, the same type hitters. But after fifty years of
> watching it I'm more convinced than ever that there
> aren't as many good hitters in the game. . . . There are
> plenty of guys with power, guys who hit the ball a long
> way, but I see so many who lack finesse, who should hit
> for average but don't. The answers are not all that hard
> to figure. They talked for years about the ball being dead.
> The ball isn't dead, the hitters are, from the neck up.

In 1975, Stan Musial, Williams's greatest contemporary from the National League, echoed similar thoughts about declining smarts in an article titled "Why the .400 Hitter Is Extinct" (in Durslag, 1975). "In order to
be successful . . . batters must have a quality that isn't too common today.
They must be able to go to the opposite field. Somehow, this art hasn't been
mastered by too many of today's players."

And lest one wrongly conclude that such thoughts circulate only
among dyspeptic old warriors, consider a journalist's opinion written in
1992, as Toronto's John Olerud made a credible bid, but fell short (Kevin
Paul Dupont in the *Boston Globe*): "Too few smart hitters. Too many guys
looking for the baddest pair of wraparound sunglasses rather than sharp-
ening the shrewdest hitting eye."

The more reasonable, and partly correct, second category—the claim
that changes in play have made batting more difficult (the Genesis Myth,
on the contrary, holds that the game is the same, but that batters have gotten soft)—includes two distinct styles of argument among its numerous
versions. I shall call these two styles "external" and "internal." The external versions maintain that commercial realities of modern baseball have
imposed new impediments upon performance.[3] This theory of "tougher

---

3. I do recognize, of course, that these claims also play into the Genesis Myth of former Elysian
fields versus modern palaces to Mammon—but my argument hinges on distinguishing the
pure Genesis Myth that batters have gotten *absolutely* worse from a more reasonable claim
that players are just as good (or better), but that batting has become *relatively* harder for some
reason.

conditions" features three common arguments, always fervently advocated when this greatest of all statistical puzzles hits the hot-stove league: too much travel within too grueling schedules; too many night games; and too much publicity and constant prying from the press (particularly when a player threatens to reach a plateau like 0.400 hitting).

The internal argument holds that aspects of the game opposed to hitting have outstripped the power of batters to compensate and respond in kind—in other words, that batters have not been able to keep up with increased sophistication in other aspects of play. This "tougher competition" theory also features three arguments (each with several subcategories)— rather obvious in this case, as representing the three institutions of baseball that might challenge good hitting:

1. Better pitching (invention of such new pitches as the slider and split-fingered fastball; the establishment of relief pitching as a specialty, with a resulting requirement for facing new and fresh arms in late innings, rather than a tired opponent seen several times before in the same game).
2. Better fielding (conversion of gloves from tiny protective coverings to much larger, ball-gobbling machines; general improvement of defense, particularly in coordination among fielders).
3. Better managing (replacement of intuitive, "seat of the pants" leadership with modern, computer-assisted assessments of strengths and weaknesses for each individual batter).

In supporting the external theory of "tougher conditions," for example, Tommy Holmes stressed the subtheme of "harder schedules" in his article "We'll Never Have Another .400 Hitter" in the February 1956 issue of *Sport* magazine:

> They [0.400 hitters of yore] started all of their single games in mid-afternoon, doubleheaders a little earlier. They never played later than sundown and usually were finished hours before dark. They did not play in the hot sunshine of one day and in the heavy damp night air the next. If they did not get the proper rest and eat

proper food at regular hours, it was nobody's fault but
their own.

For the other two subthemes, my colleague John J. Chiment of the
Boyce Thompson Institute for Plant Research at Cornell University polled
the large contingent of baseball fans in his lab, and sent me the following
in defense of the nocturnal theory (letter of April 24, 1984): "The consen-
sus at BTI favors 'Night Games' as the real problem. You just can't hit'm
if you can't see'm. Which is not to say that 'The rise of the speciality relief
pitcher' and 'Modern moral turpitude' don't have their adherents."

Finally, in June 1993, Colorado Rockies manager (and former savvy
player) Don Baylor upheld the "intrusive press" theory when his star An-
dres Galarraga and Toronto's John Olerud both exceeded 0.400 (before
their predictable decline later in the season): "Can you imagine the pressure there'd be nowadays, the press conferences that would be held after
every game? If a guy is hitting 0.400 in August?" As Olerud continued to
flirt with 0.400 in August, George Brett blamed the same source—and he
should know, for his average stood at 0.407 on August 26, 1980, while he
finished that season at 0.390. Brett remembered the journalistic assault:

> It was the same damn questions over and over and over.
> Gees, it was monotonous, and boring. In 1961, when he
> was chasing Babe Ruth [for the record of home runs in a
> season] Roger Maris lost his hair. In 1980, I got hemor-
> rhoids. I don't know what will happen to John, but I
> imagine it will be something.

The internal theory of "tougher competition" also enjoys wide support in all three major versions:

1. BETTER PITCHING.   During my lifetime as a fan, pitching has changed
more dramatically than any other aspect of the game. In my youth, during the late 1940s, most pitchers relied on curves and fastballs and they expected to work a full nine innings unless seriously shelled and tired. Relief
pitching didn't exist as a specialty; if the starter tired, the manager just put
in the next man available. Now, nearly all pitchers have expanded their

repertoire, with sliders and split-fingered fastballs as favored additions. And relief pitching has become an essential component of all good teams, with recognized subspecialties of middle relievers (good for several innings of work when starters falter) and closers (all-out throwers for a crucial final inning, day after day).

Better pitching has therefore figured prominently in attempts to explain the disappearance of 0.400 hitting. Stan Musial, for example, stated (in Durslag's article, cited previously, on "Why the .400 Hitter Is Extinct"):

> Two things have pretty much taken care of the .400 prospect. One is a thing called the slider. . . . It isn't a complicated pitch, but it's troublesome enough to take away the edge that batters used to have. A second reason is the improvement of the bullpen.

2. BETTER FIELDING. Holmes (1956, pages 37–38) cites "the tighter defenses that are rigged against the hitter today" as the primary reason for why (as his title proclaims) "We'll Never Have Another .400 Hitter." Holmes views more efficient gloves as the primary culprit (and he was writing in 1956, when gloves were downright diminutive compared with today's baskets and snares):

> Probably the sporting-goods manufacturers made an even greater contribution to curbing batting averages by producing gloves and mitts vastly superior to the ones old-timers wore. . . . The player actually did catch the ball with his hands, and his gear served to reduce the numbing impact. Now a glove is an efficient magnetic trap for the ball. . . . Today the glove, not the hand, makes the catch, with the deep pocket between the thumb and first finger doing the work.

3. BETTER MANAGING. Computers and boardroom tactics now permeate managerial staffs. Charters and number-crunchers scrutinize every swing, trying to locate a batter's weakness. Richard Hoffer (1993, page 23)

cites more "scientific" managing as the main reason for the demise of 0.400 hitting. Speaking of Williams's last success, in 1941, Hoffer writes, "He didn't have to cope with the constant charting, the defensive structure that managers routinely call into place now."

Many writers roll these conventional explanations into one large ball, and then pitch the whole kaboodle all at once. Dallas Adams, writing in the *Baseball Research Journal* for 1981 on "The Probability of the League Leader Batting .400," states:

> The commonly held view nowadays is that night ball, transcontinental travel fatigue, the widespread use of top quality relief pitchers, big ballparks, large size fielders' gloves and other factors all act to a hitter's detriment and make a .400 average a near impossibility.

Even though exhaustive repetition has enshrined these explanations as true, I believe we can conclusively debunk both versions (tougher conditions and tougher competition) of the claim that 0.400 hitting died because changes in play have made batting more difficult. The theory of tougher conditions makes no sense to me. Is transcontinental flying more tiring than those endless train trips from the East Coast to Chicago or St. Louis? Are single, air-conditioned rooms in fine hotels more conducive to exhaustion than two in a room during an August heat wave in St. Louis? Why do people continually claim that schedules are now more grueling? Modern teams play 162 games and almost no doubleheaders; during most of the century, teams played 154 games in a shorter season filled with twin bills. So who worked harder?

William Curran (1990, pages 17–18) underscores this point in writing about the conditions that a Wade Boggs (our most recent serious contender for 0.400) would have faced in the 1920s:

> First let's deprive Boggs of the services of Ted Williams as a special batting coach. Rookies of the 1920s rarely received individual instruction at any stage of their careers, and, in fact, had to fight for a chance to get into the batting cage to take a few practice swings at the ball. Next

we'll take away Wade's batting helmet and batting gloves. . . . And while we're at it, we'll have Boggs play three to five consecutive doubleheaders in the afternoon heat of September. After the games let him try to get a night's rest in St. Louis or Washington at a hotel equipped with a small room fan, if any fan at all. You get the drift.

The testimony of many players affirms the unreality of "tougher conditions" as an explanation. For example, Rod Carew, the best 0.400 prospect since Williams (and a near achiever at 0.388 in 1977), listed the litany of usual explanations and then demurred (Carew, 1979, pages 209–10):

> I don't buy much of that. I imagine that train travel was as rough as jet travel . . . and I prefer hitting at night. . . . During the day you squint a lot, and then there's a lot of stuff in the air—especially in California—and it burns your eyes. There's the glare of the sun. And in some places the artificial turf smokes up and your legs are burning. Then the perspiration during the day is running down your face. I like nighttime. You're cooler and more relaxed.

Tougher competition seems more promising because the basic facts are undoubtedly true: pitching, fielding, and managing have improved. So why shouldn't the extinction of 0.400 hitting record the relative decline of hitting as these other skills augment? All the other arguments can be refuted by the weight of their own illogic, but "tougher competition" must be tested empirically. We need to know if improvement in hitting has kept pace with opposing forces of pitching, fielding, and managing. If these three adversaries have undergone more improvement than hitting (or, even worse for batters, if hitting has stayed constant or declined as the other three factors ameliorated), then the extinction of 0.400 hitting will be well explained by "tougher competition."

But the mere fact of better pitching, fielding, and managing doesn't prove the "tougher competition" theory by itself—and for an obvious reason: hitting might have improved just as much, if not more. Why should hitting be uniquely exempt from a general betterment in all other aspects of play? Isn't it more reasonable to assume that batting has improved in concert with other factors of baseball? I shall show that general improvement in hitting has not only kept pace with betterment in other aspects of play, but that baseball has constantly fiddled with its rules to assure that major factors remain in balance. The extinction of 0.400 hitting must therefore arise from other causes.

· 8 ·

A Plausibility Argument for
General Improvement

However tempted we may be to indulge in fanciful reveries about dedication during "the good old days," the accepted notion that decline in batting skills caused the extinction of 0.400 hitting just doesn't make sense when we consider general patterns of social and sports history during the twentieth century. This context, on the contrary, almost guarantees that hitting has improved along with almost anything else we can measure at the apogee of human achievement. Consider just three of many arguments that virtually cement the case, even before we examine a single baseball statistic.

1. LARGER POOLS AND BETTER TRAINING. In 1900, 76 million people inhabited the United States, and only white men could play major league baseball. Our population has since ballooned to 249 million people (1990

census), and men of all colors and nations are welcome. Training and coaching were absent to slapdash in the past, but represent a massive industry today. Players follow rigorous and carefully calculated programs for working out (even, if not especially, during the off season, when their predecessors mostly drank beer and gained weight); they no longer risk careers and records by playing hurt. (Joe DiMaggio once told me that he was batting 0.413 with two weeks to go in the 1939 season. He caught a serious cold, which clouded his left [leading] eye, and he could not adequately see incoming pitches. The Yanks had already clinched the pennant. Any modern counterpart would sit on the bench and preserve his record; DiMaggio played to the last game and fell to 0.381, his highest seasonal average, but below the grand plateau.) No one—neither the players nor the owners—can afford to take risks and fool around today, not with star salaries in the multiple millions for careers that last but a few years at peak value. What possible argument could convince us that a smaller and more restricted pool of indifferently trained men might supply better hitters than our modern massive industry with its maximal monetary rewards? I'll bet on the larger pool, recruitment of men of all races, and better, more careful training any day.

2. SIZE.   I don't want to fall into the silly mythology of "bigger is better" (okay for a few things, like brains in the evolution of most mammalian lineages, but irrelevant for many items, like penises and automobiles). Still, *ceteris paribus* as the Romans said (all other things being equal), larger people tend to be stronger (and I say this as a short man who loved to watch Phil Rizzuto and Fred Patek). If height and weight of ballplayers have augmented through time, then (however roughly) bodily prowess should be increasing.

Pete Palmer, sabermetrician extraordinaire and editor, along with John Thorn, of *Total Baseball,* the best (and fattest) general reference book of baseball stats, sent me his chart (reproduced here as Table 1) of mean heights and weights for pitchers and batters averaged by decades. Note the remarkably steady increase through time. I cannot believe that the larger players of today are worse than their smaller counterparts of decades past.

T A B L E 1

Decadal Averages for Heights and Weights of Major League Baseball Players

| | BATTERS | | PITCHERS | |
	Height (inches)	Weight (pounds)	Height	Weight
1870s	69.1	163.7	69.1	161.1
1880s	69.6	171.6	70.2	172.7
1890s	69.8	172.1	70.6	174.1
1900s	69.9	172.6	71.5	180.7
1910s	70.3	170.5	72.1	180.7
1920s	70.4	171.2	72.0	179.8
1930s	71.1	176.8	72.6	184.8
1940s	71.4	180.3	73.0	186.5
1950s	72.0	183.0	73.1	186.1
1960s	72.2	182.7	73.6	189.3
1970s	72.3	182.3	74.1	191.0
1980s	72.5	182.9	74.5	192.2

3. RECORDS IN OTHER SPORTS.   All major baseball records are relative—that is, they assess performance against other players in an adversarial role—not absolute as measured by personal achievement, and counted, weighed, or timed by a stopwatch. A 0.400 batting average records degree of relative success against pitchers, whereas a four-minute mile, a nineteen-foot pole vault, or a 250-pound lift is unvarnished *you* against an unchanging outer world.

Improvements in relative records are ambiguous in permitting several possible (and some diametrically opposed) interpretations: rising batting averages might mean that hitting has improved, but the same increase might also signify that batting has gotten worse while pitching has deteriorated even more sharply (leading to relative advantages for hitters as their absolute skills eroded).

Absolute records, however, have clearer meaning. If leading sprinters are running quicker and vaulters jumping higher . . . well, then they are performing their art better. What else can we say? The breaking of records doesn't tell us why modern athletes are doing better—and a range of diverse reasons might be cited, from better training, better understanding of human physiology, new techniques (the Beamon back flop), to new equipment (fiberglass poles and the immediate, dramatic rise in record heights for pole vaulting)—but I don't think that we can deny the fact of improvement.

Therefore, since the relative records of baseball must be ambiguous in their causes, we should study the absolute records of related sports. If most absolute records have been improving, then shouldn't we assume that athletic prowess has risen in baseball as well? Wouldn't we be denying a general pattern and creating an implausible, ad hoc theory if we attributed the extinction of 0.400 hitting to a decline in batting skills? Shouldn't we be searching for a theory that can interpret the death of 0.400 hitting as a consequence of generally superior athleticism—thus making this most interesting and widely discussed trend in the history of baseball statistics consistent with the pattern and history of almost every other sport?

I don't want to worry a well-understood subject to death, and I don't want to bore you with endless documentation of well-known phenomena. Surely all sports fans recognize the pervasive pattern of improvement in absolute records through time. The first modern Olympic marathon champ, Spiridon Loues, took almost three hours in 1896; more recent winners are nearly down to two. The allure of the four-minute mile challenged runners for decades, while Paavo Nurmi's enticing 4.01 held from 1941 until Roger Bannister's great moment on May 6, 1954. Now, most of the best runners routinely break four minutes nearly every time. By 1972, for the 100-meter freestyle, and by 1964 for the 400 meters, the best women swimmers eclipsed the Olympic records of the 1920s and 1930s, set by the two great Tarzans (both played the role in movies) Buster Crabbe and Johnny Weismuller. I will let one chart stand for the generality, based on

FIGURE 12

The steadily decreasing record time for men in the Boston Marathon. Dots are five-year averages (my calculation).

data closest to hand for the most famous of local races at my workplace— the Boston Marathon (see Figure 12). The general pattern is clear, and the few anomalies record changes in distance (the "standard" 26 miles 385 yards has prevailed in most years, but early winners, from 1897 to 1923, ran only 24 miles 1,232 yards for their longer times; with a rise to 26 miles 209 yards from 1924 to 1926; the standard distance from 1927 to 1952; and a shortened 25 miles 958 yards from 1953 to 1956, until reestablishment of the standard distance in 1957).

For almost every sport, the improvement in absolute records follows a definite pattern with presumed causes central to my developing argument about 0.400 hitting. Improvement does not follow a linear path of constant rate. Rather, times and records fall more rapidly early in the sequence and then slow markedly, sometimes reaching a plateau of no further advance (or of minutest measurable increments from old records). In other words, athletes eventually encounter some kind of barrier to future progress, and records stabilize (or at least slow markedly in their frequency and amount of improvement). Statisticians call such a barrier an asymptote; vernacular language might speak of a limit. In the terminology of this book, athletes reach a "right wall" that stymies future improvement.

Since we are considering the world's best performers in these calculations, the probable reasons for such limits or walls should be readily apparent. After all, bodies are physical devices, subject to constraints upon performance set by size, physiology, and the mechanics of muscles and joints. No one will argue that curves of improvement can be extrapolated forever—or else runners would eventually complete the mile in nothing flat (and, finally, in negative time), and pole vaulters would truly match a gentleman of legend and leap tall buildings in a single bound.

We can best test the proposition that physical limits (or right walls) cause the slowing and plateauing of improvements by comparing curves for athletes operating near the extremes in human capacity with performers who probably retain much room for further advance. What conditions might place people far from the right wall, and therefore endow them with great scope for improvement? Consider some potential examples: new sports where athletes have not yet figured out optimal procedures; new categories of people recently admitted to old sports; records for amateur play. As an example, the Boston Marathon was opened to women only in 1972. Note how much more rapidly women have improved than men from their beginning to the present (Figure 13).

We may generalize this principle by setting up a hierarchy of decreasing improvement (also a ranking of increasing worth in the value systems of some rather old-fashioned and well-heeled folks): women, men, and horses. Winning times for major horse races have improved, but ever so slightly over long intervals. For example, between 1840 and 1980, thoroughbred horses in the three great English races of St. Leger, Oaks, and the Derby have shaved twelve, twenty, and eighteen seconds off record times, for a minuscule gain of 0.4 to 0.8 percent per generation (Eckhardt et al., 1988). These gains are tiny even when compared with the other great arena of breeding in domesticated animals: improvement of livestock, where gains of 1 to 3 percent per year are often achieved for features of economic importance.

This limited improvement makes perfect and predictable sense. Thoroughbreds have been rigorously raised from a limited stock for more than two hundred years. Stakes could not be higher, as the slightest improvement may be worth millions. More effort has gone into betterment of this breed than into almost any other biological endeavor of economic impor-

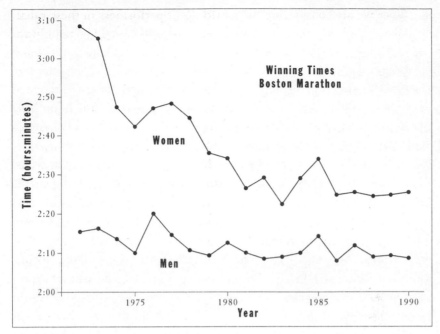

FIGURE 13

Women's records decline precipitously while men's records remain relatively stable between 1970 and 1980 in the Boston Marathon. Since 1980 both records have changed little. Dots are actual winning times for each year.

tance to humans. We might therefore suppose that the best thoroughbreds have long resided at the genetic right wall for the breed, and that future improvements will be negligible to slow. But since (thank God) we have not yet reached brave new world, we do not breed humans for optimized physical performance, and records for people should therefore show more flexibility—for we have no purposeful purebreds at our right walls.

In most popular and established men's events, we note the pattern of rapid initial improvement followed by flattening of the curve.[4] Exceptions may be found in such events as the marathon, where length and complexity provide great "play" for experimenting with new strategies, and where re-

---

4. All bets are off when fundamentally new equipment or procedures enter the field, as in the fiberglass pole, or (God forbid) the aluminum bat, which (we may hope and pray) will never darken the doorstep of major league dugouts. Such innovations will produce sudden blips in curves of improvement. In fact, such innovations are usually better treated statistically as the beginning points of new curves.

cent surges in popularity have brought large increases in prestige and participation. (Note that the curve of improvement for the Boston Marathon has remained virtually linear for men, and did not slow before 1990—though the pattern may now be shifting into the usual mode as the world's best runners now compete and improvements begin to abate.)

Many commentators have noted that most women's records are both falling faster than men's for the same event, and are not yet flattening, but maintaining a linear pace of improvement. Interestingly (see Whipp and Ward, 1992), most men's running events (200 to 10,000 meters) have improved in the same range of rates regardless of the event's total distance—5.69 to 7.57 meters per minute improvement per decade. (Improvement in the marathon has been greater, at 9.18 meters per minute per decade, thus supporting my claim that this event remains "immature" and still in the category of potentially linear improvement—that is, not near the right wall.) But for women in the same events, rates of improvement run from 14.04 to 17.86 minutes per meter per decade (with a whopping 37.75 meters for the marathon).

These findings have led to all manner of speculation, some rather silly. For example, Whipp and Ward (1992) just extrapolate their curves and then defend the conclusion that women will eventually outrun men in most events, and rather soon for some. (The extrapolated curves for the marathon, for example, cross in 1998 when women should beat men by this argument.)

But extrapolation is a dangerous, generally invalid, and often foolish game. After all, as I said before, extrapolate the linear curve far enough and all distances will be run in zero and then in negative time. (False extrapolation also produces the irresponsible figures often cited for growth of human populations—in a few centuries, for example, humans will form a solid mass equal to the volume of the earth and no escape into outer space will be possible because the rate of increase will cause the diameter of this human sphere to grow at greater than the speed of light, which, as Einstein taught us, sets an upper bound upon rapidity of motion.) Clearly we will never run in negative time, nor will our sphere of solid humanity expand at light speed. Limits or right walls will be reached, and rates of increase will first slow and eventually stop.

Women may outclass men in certain events like ultra-long-distance swimming, where buoyancy and fat distribution favor women's physiques and endurances over men's (women already hold the absolute record for the English Channel and Catalina Island swims). The marathon may also be a possibility. But I doubt that women will ever capture either the 100-meter dash or the heavyweight lifting records. (Many women will always beat most men in any particular event—most women can beat me in virtually anything physical. But remember that we are talking of world records among the very best performers—and here the biomechanics of different construction will play a determining role.)

The basic reason for more rapid gains (and less curve flattening) in women's events seems clear. Sexism is the culprit, and happy reversals of these older injustices the reward. Most of these events have been opened to women only recently. Women have been brought into the world of professionalism, intense training, and stiff competition only in the last few years. Women, not so long ago (and still now for so many), were socialized to regard athletic performance as debarred to their gender—and many of the great women performers of the past, Babe Didrikson in particular, suffered the onus of wide dismissal as overly masculine. In other words, most women's curves are now near the beginning of the sequence— in the early stages of rapid and linear improvement. These curves will flatten as women reach their own right walls—and only then will we know true equality of opportunity. Until then, the steep and linear improvement curves of women's sports stand as a testimony to our past and present inequities.

·9·

0.400 Hitting Dies as the Right Tail Shrinks

Granting the foregoing argument that hitting must be improving in some absolute sense as the best athletes first rush, and then creep, toward the right wall of biomechanical limits on human performance, only one traditional explanation remains unrefuted for viewing the extinction of 0.400 hitting as the deterioration of something at bat—the possibility that, while hitting has improved, other opposing activities (pitching and fielding) have gotten better, even faster, leading to a *relative* decline in batting performance.

This last holdout of traditionalism fails the simplest and most obvious test of possible validity. If pitching and fielding have slowly won an upper hand over hitting, we should be able to measure this effect as a general decline in batting averages through the twentieth-century history of baseball. If mean batting averages have fallen with time, as pitching and fielding assert increasing domination, then the best hitters (the 0.400 men of yore) get dragged down along with the masses—that is, if the mean batting average were once 0.280, then a best of over 0.400 makes sense as an upper bound, but if the mean has now fallen to, say, 0.230, then 0.400 might stand too far from this declining mean for even the best to reach.

This entirely sensible explanation fails because, in fact, the mean batting average for everyday players has been rock-stable throughout our century (with interesting exceptions, discussed later, that prove the rule). Table 2 (page 102) presents decadal mean batting averages for all regular players in both leagues during the twentieth century. (I included only those players who averaged more than two at-bats per game for the entire season, thus eliminating weak-hitting pitchers and second-stringers hired for their skills in fielding or running.)[5] The mean batting average began at about 0.260, and has remained there throughout our century. (The sustained, though temporary, rise in the 1920s and 1930s presents a single and sensible exception, for reasons soon to come, but cannot provide an explanation for the subsequent decline of 0.400 hitting for two reasons: first, the greatest age of 0.400 hitting occurred before then, while averages stood at their usual level; second, not a soul hit over 0.400 throughout the 1930s, despite the high league means—I include Bill Terry's 0.401 of 1930 itself in the 1920s calculation.) Thus our paradox only deepens: 0.400 hitting disappeared in the face of preserved constancy in average performance. Why should the best be trimmed, while ordinary Joes continue to perform as ever before? We must conclude that the extinction of 0.400 hitting does not reflect a general decline in batting prowess, either absolute or relative.

When issues reach impasses of this sort, we usually need to find an exit by reformulating the question—and reentering the field by another door. In this case, and following the general theme of my book, I suggest that

---

5. The recent disparity between the two leagues records, in large part, the introduction of the "designated hitter" to the American League alone—a permanent "pinch hitter" for the pitcher. His substitution for the pitcher doesn't affect the decadal average per se, because I don't include pitchers in this calculation. But the designated hitter still provokes a small general rise in the American League mean by introducing another good bat into the lineup, whereas the National League retains more relatively poor hitters in the bottom part of the order. Nonetheless, I remain an adamant opponent of the DH rule—the one vital subject in our culture that permits no middle ground. You gotta either love it or hate it!

we have been committing the deepest of all errors from the start of our long-standing debate about the decline of 0.400 hitting. We have erred—unconsciously to be sure, for we never considered an alternative—by treating "0.400 hitting" as a discrete and definable "thing," as an entity whose disappearance requires a special explanation. But 0.400 hitting is not an item like "Joe DiMaggio's favorite bat," or even a separately definable class of objects like "improved fielders' gloves of the 1990s." We should take a hint from the guiding theme of this book: the variation of a "full house" or complete system should be treated as the most compelling "basic" reality; averages and extreme values (as abstractions and unrepresentative instances respectively) often provide only partial, if not downright misleading, views of a totality's behavior.

Hitting 0.400 is not an item or entity, a thing in itself. Each regular player compiles a personal batting average, and the totality of these averages may be depicted as a conventional frequency distribution, or bell curve. This distribution includes two tails for worst and best performances—and the tails are intrinsic parts of the full house, not detachable items with their own individuality. (Even if you could rip a tail off, where would you make the break? The tails grade insensibly into the larger center of the distribution.) In this appropriately enlarged perspective, 0.400 hitting is the right tail of the full distribution of batting averages for all players, not in any sense a definable or detachable "thing unto itself." In fact, our propensity for recognizing such a category at all only arises as a psychological outcome of our quirky propensity for dividing smooth continua at numbers that sound "even" or "euphonious"—witness our excitement about the coming millennial transition, though the year 2000 promises no astronomical or cosmic difference from 1999 (see Gould, 1996, essay 2).

When we view 0.400 hitting properly as the right tail of a bell curve for all batting averages, then an entirely new form of explanation becomes possible for the first time. Bell curves can expand or contract as amounts of variation wax or wane. Suppose that a frequency distribution maintains the same mean value, but that variation diminishes symmetrically, with more individual measures near the mean and fewer at both the right and left tails. In that case, 0.400 hitting might then disappear entirely, while the mean batting average remained stable—but the cause would then reside in whatever set of reasons produced the shrinkage of variation around a constant mean. This different geometrical picture for the disappearance of 0.400 hitting does not specify a reason, but the new model forces us to reconsider the entire issue—for I can't think of a reason why a general shrinkage of variation should record the worsening of anything. In fact, the opposite might be true: perhaps a general shrinkage of variation reflects improvement in the state of baseball. At the very least, this reformulation weans us from traditional, locked-in, and unproductive modes of explanation—in this case the "certainty" that extinction of 0.400 hitting must be recording a trend in the degeneration of batting skills. We are now free to consider new explanations: Why should variation be shrinking? Does shrinking record improvement or degeneration (or neither)—and, if so, of what?

Does this alternate explanation work? I have already documented the first part of the claim—preservation of relatively constant mean batting averages through time (see Table 2). But what about the second component? Has variation been shrinking symmetrically about this mean value during the history of twentieth-century baseball? Let me first demonstrate that mean batting averages have been stabilized by an active effort of rulemakers—for natural shrinkage about a purposely fixed point presents an appealing picture that, in my view, establishes our best argument for viewing 0.400 hitting as a predictable and inevitable consequence of general improvement in play.

Figure 14 presents mean batting averages for all regular players in both leagues year by year (the National League began in 1876, the American League in 1901). Note the numerous excursions in both directions, but invariable returns to the general 0.260 level. This average level has been actively maintained by judicious modification of the rules whenever hitting or pitching gained a temporary upper hand and threatened to disrupt the saintly stability of our national pastime. Consider all the major fluctuations:

After beginning at the "proper" balance, averages began to drift down, reaching the 0.240s during the late 1880s and early 1890s. In response, and in the last major change ever introduced in the fundamental structure of baseball (number 1 on Figure 14), the pitching mound retreated to its current distance of sixty feet six inches from the plate during the 1893 season. (The mound had begun at forty-five feet from the plate, with pitchers de-

TABLE 2

LEAGUE AVERAGES FOR THE TWENTIETH CENTURY, BY DECADES

	AMERICAN LEAGUE	NATIONAL LEAGUE
1901–1910	.251	.253
1911–1920	.259	.257
1921–1930	.286	.288
1931–1940	.279	.272
1941–1950	.260	.260
1951–1960	.257	.260
1961–1970	.245	.253
1971–1980	.258	.256
1981–1990	.262	.254

livering the ball underhand, and had migrated steadily back during base-ball's early days—the reason for limited utility of nineteenth-century sta-tistics in these calculations.) Unsurprisingly, hitters responded with their best year ever. The mean batting average soared to 0.307 in 1894, and re-mained high until 1901 (number 2 on Figure 14), when adoption of the foul-strike rule forced a rapid decline to propriety (foul balls had not pre-viously been counted for strikes one and two). Averages remained anom-alously low until introduction of the cork-centered ball prompted an abrupt rise in 1911 (number 3 in Figure 14). Pitchers quickly accommo-dated, and averages returned to their proper 0.260 level as the decade ad-vanced.

The long excursion (number 4 on Figure 14), nearly twenty years of high hitting during the 1920s and 1930s, represents the one extended ex-ception to a pattern of long stability interrupted by quick blips—and the fascinating circumstances and putative reasons have long been debated by all serious fans. In 1919, Babe Ruth hit a wildly unprecedented twenty-

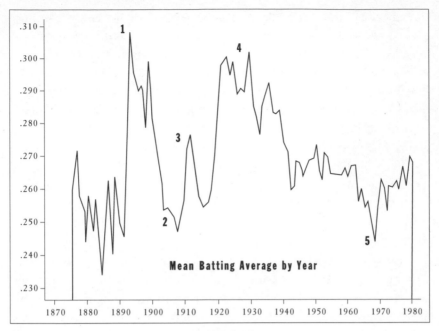

FIGURE 14

The mean batting average for regular major league players has remained quite steady at about 0.260 during the entire history of major league baseball. The few exceptions can be explained and were "corrected" by judicious changes in the rules. Averages rose after the pitching mound was moved back *(1)*; declined after adoption of the foul-strike rule *(2)*; rose again after the introduction of the cork-centered ball *(3)*, and then again during the 1920s and '30s *(4)*. The decline in the 1960s *(5)* was reversed in 1969 by lowering the pitching mound and decreasing the strike zone.

nine homers, more than most entire teams had garnered in full seasons before; then, in 1920, he nearly doubled the total, to fifty-four. At all other times, the moguls of baseball would have reacted strongly to this unseemly change and would, no doubt, have reined in these Ruthian tendencies by some judicious change of rules. But 1920 represented the crux of a unique threat in the history of baseball. Several members of the 1919 Chicago White Sox (the contingent later known as the Black Sox), including the great 0.400 hitter Shoeless Joe Jackson, had accepted money from a gambling ring to throw the World Series of 1919. The resulting revelations almost destroyed professional baseball, and attendance declined precipitously during the 1920 season. The owners (whose pervasive stinginess had set the context that encouraged such admittedly dishonest and indefensible behavior) turned to Ruth as a *deus ex machina*. His new style of play packed in the crowds, and owners, for once, went with the flow and allowed the game to change radically. Scrappy, one-run-at-a-time, anyway-possible, savvy-baserunning, pitcher's baseball became a style of the past (much to Ty Cobb's permanent disgust); big offense and swinging for the fences became *de rigueur*. Mean batting averages rose abruptly and remained high for twenty years, even breaking 0.300 for the second (and only other) time in 1930.

But why were Ruth and other hitters able to perform so differently when circumstances encouraged such a change? Traditional wisdom—it is ever so, as we search for the "technological fix"—attributes this long plateau of exalted batting averages to introduction of a "lively ball." But Bill James, baseball's greatest sabermetrician, argues (in his *Historical Baseball Abstract,* Villard Books, 1986) that no major fiddling with baseballs in 1920 can be proven. James suspects that balls did not change substantially, and that rising batting averages can be attributed to alterations in rules and attitudes that imposed multiple and simultaneous impediments upon pitching, thus upsetting the traditional balance for twenty years. All changes in practice favored hitters. Trick pitches were banned, and hurlers who had previously scuffed, shined, and spat on balls with abandon now had to hide their antics. Umpires began to supply shiny new balls whenever the slightest scuff or spot appeared. Soft, scratched, and darkened balls had previously remained in play as long as possible—fans even threw back foul balls (!), as they do today in Japan, except for home runs. James argues that the immediate replacement of soft and discolored by hard and shiny balls would do as much for improved hitting as any supposedly new construction of a more tightly wound, livelier ball.

In any case, averages returned to their conventional level in the 1940s as the war years siphoned off the best in all categories. Since then, only one interesting excursion has occurred (number 5 in Figure 14)—another fine illustration of the general principle, and recent enough to be well remembered by millions of fans. For reasons never determined, batting averages declined steadily throughout the 1960s, reaching a nadir in the great pitchers' year of 1968, when Carl Yastrzemski won the American League batting title with a minimal 0.301, and Bob Gibson set his astonishing, off-scale record of a 1.12 earned run average (see page 127 for more on Gibson). So what did the moguls do? They changed the rules, of course—this time by lowering the pitching mound and decreasing the strike zone. In 1969, mean batting averages returned to their usual level—and have remained there ever since.

I do not believe that rulemakers sit down with pencil and paper, trying to divine a change that will bring mean batting averages back to an ideal. Rather, the prevailing powers have a sense of what constitutes proper balance between hitting and pitching, and they jiggle minor factors accordingly (height of mound, size of strike zone, permissible and impermissible alterations of the bat, including pine tar and corking)—in order to assure stability within a system that has not experienced a single change of fundamental rules and standards for more than a century.

But the rulemakers do not (and probably cannot) control amounts of variation around their roughly stabilized mean. I therefore set out to test my hypothesis—based on the alternate construction of reality as the full house of "variation in a system" rather than "a thing moving somewhere"—that 0.400 hitting (as the right tail in a system of variation rather than a separable thing-in-itself) might have disappeared as a consequence of shrinking variation around this stable mean.

I did my first study "on the cheap" when I was recovering from serious illness in the early 1980s (see chapter 4). I propped myself up in bed with the only book in common use that is thicker than the Manhattan telephone directory—*The Baseball Encyclopedia* (New York, Macmillan). I decided to treat the mean batting average for the five best and five worst players in each year as an acceptable measure of achievement at the right and left tails of the bell curve for batting averages. I then calculated the difference between these five highest and the league average (and also between the five lowest and the league average) for each year since the beginning of major league baseball, in 1876. If the difference between best and average (and worst and average) declines through time, then we will have a rough measurement for the shrinkage of variation.

The five best are easily identified, for the *Encyclopedia* lists them in yearly tables of highest achievement. But nobody bothers to memorialize the five worst, so I had to go through the rosters, man by man, looking for the five lowest averages among regular players with at least two at-bats per game over a full season. I present the results in Figure 15—a clear con-

firmation of my hypothesis, as variation shrinks systematically and symmetrically, bringing both right and left tails ever closer to the stable mean
through time. Thus, the disappearance of 0.400 hitting occurred because
the bell curve for batting averages has become skinnier over the years, as
extreme values at both right and left tails of the distribution get trimmed
and shaved. To understand the extinction of 0.400 hitting, we must ask
why variation declined in this particular pattern.

Several years later I redid the study by a better, albeit far more labo-
rious, method of calculating the conventional measure of total variation—
the standard deviation—for all regular players in each year (three weeks
at the computer for my research assistant—and did he ever relish the
break from measuring snails!—rather than several enjoyable personal
hours propped up in bed with the *Baseball Encyclopedia*).

The standard deviation is a statistician's basic measure of variation.
The calculated value for each year records the spread of the entire bell
curve, measured (roughly) as the average departure of players from the
mean—thus giving us, in a single number, our best assessment of the full
range of variation. To compute the standard deviation, you take (in this
case) each individual batting average and subtract from it the league average for that year. You then square each value (multiply it by itself) in
order to eliminate negative numbers for batting averages below the mean
(for a negative times a negative yields a positive number). You then add
up all these values and divide them by the total number of players, giving
an average squared deviation of individual players from the mean. Finally,
you take the square root of this number to obtain the average, or standard,
deviation itself. The higher the value of the standard deviation, the more
extensive, or spread out, the variation.[6]

Calculation by standard deviation gives a more detailed account of the
shrinkage of variation in batting averages through time—see Figure 16,

---

6. I referred to my first method as working "on the cheap" because five-highest and five-lowest
represents a quicker and dirtier calculation than the full standard deviation of all players.
But I knew that this shortcut would provide a good surrogate for the more accurate standard deviation because standard deviations are particularly sensitive to values farthest from
the mean—a consequence of squaring the deviation of each player from the mean at one point
in the calculation. Since my quick-and-dirty method relied entirely on values farthest from
the mean, I knew that it would correlate closely with the standard deviation.

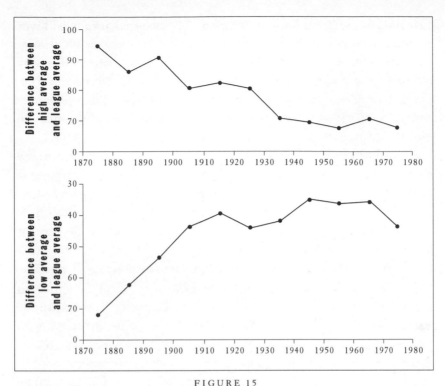

FIGURE 15

Declining differences between highest and lowest batting averages and league means throughout the history of baseball.

which plots the changes in standard deviation year by year, with no averaging over decades or other intervals. My general hypothesis is confirmed again: variation decreases steadily through time, leading to the disappearance of 0.400 hitting as a consequence of shrinkage at the right tail of the distribution. But, using this preferable, and more powerful, method of standard deviations, we can discern some confirming subtleties in the pattern of decrease that our earlier analysis missed. We note in particular that, while standard deviations have been dropping steadily and irreversibly, the decline itself has decelerated over the years as baseball has stabilized—rapidly during the nineteenth century, more slowly during the twentieth, and reaching a plateau by about 1940.

Please pardon a bit of crowing, but I was stunned and delighted (beyond all measure) by the elegance and clarity of this result. I knew from my previous analysis what the general pattern would show, but I never dreamed that the decline of variation would be so regular, so devoid of exception or anomaly for even a single year, so unvarying that we could even pick out such subtleties as a deceleration in decline. I have spent my entire professional career studying such statistical distributions, and I know how rarely one obtains such clean results in better-behaved data of controlled experiments or natural growth in simple systems. We usually encounter some glitch, some anomaly, some funny years. But the decline of standard deviations for batting averages is so regular that the pattern of Figure 16 looks like a plot for a law of nature.

I find the regularity all the more remarkable because the graph of mean batting averages themselves through time (Figure 14) shows all the noise and fluctuation expected in natural systems. These mean batting averages have frequently been manipulated by the rulemakers of baseball to maintain a general constancy, while no one has tried to monkey with the standard deviations. Nonetheless, while mean batting averages go up and down to follow the whims of history and the vagaries of invention, the standard deviation has marched steadily down at a decreasing pace, apparently disturbed by nothing of note, apparently following some interesting rule or general principle in the behavior of systems—a principle that should provide a solution to the classic dilemma of why 0.400 hitting has disappeared.

The details of Figure 16 are impressive in their exceptionless regularity. All four beginning years of the 1870s feature high values of standard deviations greater than 0.050, while the last reading in excess of 0.050 occurs in 1886. Values between 0.04 and 0.05 characterize the remainder of the nineteenth century, with three years just below at 0.038 to 0.040. But the last reading in excess of 0.040 occurs in 1911. Subsequently, decline within the 0.03-to-0.04 range shows the same precision of detail in unreversed decrease over many years. The last reading as high as 0.037 occurs in 1937, and of 0.035 in 1941. Only two years have exceeded 0.034 since 1957. Between 1942 and 1980, values remained entirely within the restricted range of 0.0285 to 0.0343. I had thought that at least one unusual year would upset the pattern, that at least one nineteenth-century value would reach a late-twentieth-century low, or one recent year soar to a nineteenth-century high—but nothing of the sort occurs. All yearly mea-

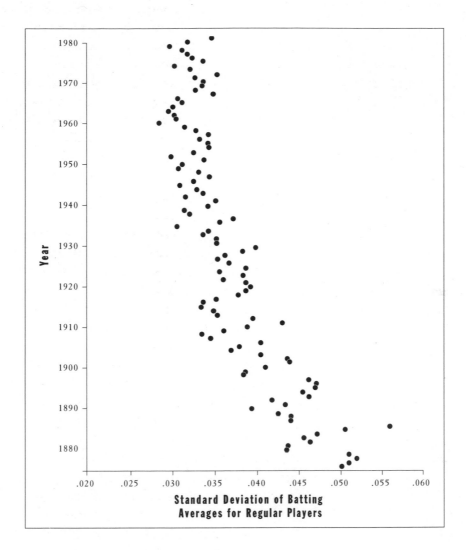

FIGURE 16

Standard deviation of batting averages for all full-time players by year for the first 100 years of professional baseball. Note the regular decline.

sures from 1906 back to the beginning of major league baseball are higher than every reading from 1938 to 1980. We find no overlap at all. Speaking as an old statistical trouper, I can assure you that this pattern represents regularity with a vengeance. This analysis has uncovered something general, something beyond the peculiarity of an idiosyncratic system, some rule or principle that should help us to understand why 0.400 hitting has become extinct in baseball.

·10·

Why the Death of 0.400 Hitting Records Improvement of Play

So far I have only demonstrated a pattern based on unconventional concepts and pictures. I have not yet proposed an explanation. I have proposed that 0.400 hitting be reconceptualized as an inextricable segment in a full house of variation—as the right tail of the bell curve of batting averages—and not as a self-contained entity whose disappearance must record the degeneration of batting in some form or other.

In this different model and picture, 0.400 hitting disappears as a consequence of shrinking variation around a stable mean batting average. The shrinkage is so exceptionless, so apparently lawlike in its regularity, that we must be discerning something general about the behavior of systems through time.

Why should such a shrinkage of variation record the worsening of anything? The final and explanatory step in my argument must proceed beyond the statistical analysis of batting averages. We must consider both the nature of baseball as a system, and some general properties of systems that enjoy long persistence with no major changes in procedures and behaviors. I therefore devote this section to reasons for celebrating the loss of 0.400 hitting as a mark of better baseball.

Two arguments, and supporting data, convince me that shrinkage of variation (with consequent disappearance of 0.400 hitting) must be measuring a general improvement of play. The two formulations sound quite dissimilar at first, but really represent different facets of a single argument.

1. *Complex systems improve when the best performers play by the same rules over extended periods of time. As systems improve, they equilibrate and variation decreases.* No other major American sport permits such an analysis, for all others have changed their fundamental rules too often and too recently. As a teenager, I played basketball without the twenty-four-second rule. My father played with a center jump after each basket. His father (had he been either inclined or acculturated) would have brought the ball downcourt with a two-handed dribble. And Mr. Naismith's boys threw the ball into a peach basket. While the peach basket still hung in the 1890s, baseball made its last major change in procedure (as discussed in the last chapter) by moving the pitcher's mound back to the current distance of sixty feet six inches.

But constant rules don't imply unchanging practices. (In the last chapter I discussed the numerous fiddlings and jigglings imposed by rulemakers to keep pitching and hitting in balance.) Dedicated performers are constantly watching, thinking, and struggling for ways to twiddle or manipulate the system in order to gain a legitimate edge (new techniques for hitting a curve, for gobbling up a ground ball, for gyrating in a windup to fool the batter). Word spreads, and these minor discoveries begin to pervade the system. The net result through time must inevitably encourage an ever-closer approach to optimal performance in all aspects of play—combined with ever-decreasing variation in modes of procedure.

Baseball was feeling its juvenile way during the early days of major league play. The basic rules of the 1890s are still our rules, but scores of subtleties hadn't yet been invented or developed. Rough edges careered out in all directions from a stable center. To cite just a few examples (taken from Bill James's *Historical Baseball Abstract*): pitchers only began to cover first base in the 1890s. During the same decade, Brooklyn developed the cutoff play, while the Boston Beaneaters invented the hit-and-run, and signals from runner to batter. Gloves were a joke in these early days—just a bit of leather over the hand, not today's baskets for trapping balls. As a fine symbol of broader tolerance and variation, the 1896 Philadelphia Phillies actually experimented for seventy-three games with a lefty shortstop. Unsurprisingly, traditional wisdom applied. He stank—turning in the worst fielding average with the fewest assists among all regular shortstops in the league.

In baseball's youth, styles of play had not become sufficiently regular and optimized to foil the accomplishments of the very best. Wee Willie Keeler could "hit 'em where they ain't" (his motto), and compile a 0.432 batting average in 1897, because fielders didn't yet know where they should be. Slowly, by long distillation of experience, players moved toward optimal methods of positioning, fielding, pitching, and batting—and variation inevitably declined. The best now meet an opposition too finely honed to its own perfection to permit the extremes of accomplishment that characterized a more casual and experimental age. We cannot explain the disappearance of 0.400 hitting simply by saying (however true) that managers invented relief pitching, while pitchers invented the slider—for such traditional explanations abstract 0.400 hitting as an independent phenomenon and view its extinction as the chief sign of a trend to deterioration in batting. Rather, hitting has improved along with all other aspects of play as the entire game sharpened its standards, narrowed its ranges of tolerance, and therefore limited variation in performance as all parts of the game climbed a broader-based hill toward a much narrower pinnacle.

Consider the predicament of a modern Wade Boggs, Tony Gwynn, Rod Carew, or George Brett. Can anyone truly believe that these great hitters are worse than Wee Willie Keeler (at five feet four and a half inches and 140 pounds), Ty Cobb, or Rogers Hornsby? Every pitch is now charted, every hit mapped to the nearest square inch. Fielding and relaying have improved dramatically. Fresh and rested pitching arms must be faced in the late innings; fielders scoop up grounders in gloves as big as a brontosaurus's footprint. Relative to the right wall of human limitation, Tony Gwynn and Wee Willie Keeler must stand in the same place—just a few inches from theoretical perfection (the best that human muscles and bones can do). But average play has so crept up upon Gwynn that he lacks the space for taking advantage of suboptimality in others. All these general improvements must rob great batters of ten to twenty hits a year—a bonus that would be more than enough to convert any of the great modern batters into 0.400 hitters.

I have formulated the argument parochially in the terms and personnel of baseball. But I feel confident that I am describing a general property of systems composed of individual units competing with one another under stable rules and for prizes of victory. Individual players struggle to find means for improvement—up to limits imposed by balances of competition and mechanical properties of materials—and their discoveries accumulate within the system, leading to general gains toward an optimum. As the system nears this narrow pinnacle, variation must decrease— for only the very best can now enter, while their predecessors have slowly, by trial and error, discovered better procedures that now cannot be substantially improved. When someone discovers a truly superior way, everyone else copies and variation diminishes.

Thus I suspect that similar reasons (along with a good dollop of historical happenstance) govern the uniformity of automotive settling upon internal combustion engines from a wider set of initial possibilities including steam and electric power; the standardization of business practices; the reduction of life's initial multicellular animal diversity to just a handful of major phyla (see Gould, 1989); and the disappearance of 0.400 hitting in baseball as variation shrinks symmetrically around a stable mean batting average.

In the good old days of greater variation and poorer play, you could get a job for "good field, no hit"—but no longer as the game improved and the pool of applicants widened. So the left tail shriveled up and moved toward the mean. In those same legendary days, the very best hitters could take advantage of a sloppier system that had not yet discovered optimalities of opposing activities in fielding and pitching. Our modern best hitters are just as good and probably better, but average pitching and fielding have so improved that the truly superb cannot soar so far above the ordinary. Therefore the right tail shriveled up and also moved toward the mean.

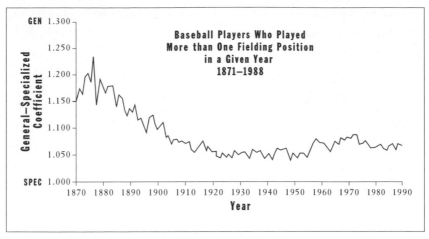

FIGURE 17

Increasing specialization as shown by decline in the number of players who fielded more than one position in a given year.

I first published these ideas in the initial issue of the revived *Vanity Fair* in March 1983. To my gratification, several fellow sabermetricians became intrigued and took up the challenge to test my ideas with other sources of baseball data. The results have been most gratifying. In particular, my colleagues have provided good examples of the two most important predictions made by models for general improvement marked by decreasing variation.

Specialization and division of labor. Ever since Adam Smith began *The Wealth of Nations* with his famous example of pinmaking, specialization and division of labor have been viewed as the major criteria of increasing efficiency and approach to optimality. In their paper "On the tendency toward increasing specialization following the inception of a complex system—professional baseball 1871–1988," John Fellows, Pete Palmer, and Steve Mann plotted the number of major leaguers who played more than one fielding position in a single season. Note (see Figure 17) the steady decrease and subsequent stabilization, a pattern much like the decelerating decline of standard deviations in Figure 16—though in this case measuring the increase of specialization through baseball's history (I do not know why values rose slightly in the 1960s, though to nowhere near the high levels of baseball's early history).

Decreasing variation. My colleagues Sangit Chatterjee and Mustafa Yilmaz of the College of Business Administration at Northeastern University (baseball does provide some wonderful cohesion amid our diversity) wrote an article on "Parity in baseball: stability of evolving systems." In searching for an example even more general than shrinking variation in batting averages, Chatterjee and Yilmaz reasoned that if general play has improved, with less variation among a group of consistently better players, then disparity among teams should also decrease—that is, the difference between the best and worst clubs should decline because all teams can now fill their rosters with enough good players, leading to greater equalization through time. The authors therefore plotted the standard deviation in seasonal winning percentage from the beginning of major league baseball to the present. Figure 18 shows a steady fall in standard deviation, indicating a decreasing difference between the best and worst teams through the history of play.[7]

2. *As play improves and bell curves march toward the right wall, variation must shrink at the right tail.* I discussed the notion of "walls" in chapter 4— upper and lower limits to variation imposed by laws of nature, structure of materials, etc. (There I illustrated a minimal left wall in the story of my medical history—an obvious and logical lower bound of zero time between diagnosis and death from the same disease. Part Four will focus upon a left wall of minimal complexity for life—for nothing much simpler than a bacterial cell could be preserved in the fossil record.) We would all, I think, accept the notion that a "right wall" must exist for human achievement. We cannot, after all, perform beyond the limits of what human bone and muscle can accomplish; no man will ever outpace a cheetah or a finch. We would also, I assume, acknowledge that some extraordinary people,

---

7. These statistics can also be broken down to yield finer patterns that validate the hypothesis. The National League began in 1876, the American in 1901. Since the hypothesis holds that systems equilibrate through time by decelerating decrease in variation, we might predict that, from 1901 to 1930, when the American League was new but the National already in middle age, variation in American League records should decrease more rapidly than comparable measures in the National League. This pattern does indeed emerge, both for standard deviations of batting averages in my calculations, and for the history of differences in best versus worst teams in the data of Chatterjee and Yilmaz.

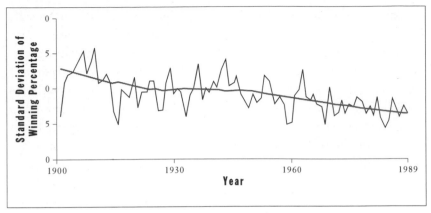

FIGURE 18

Decline in the standard deviation of winning percentage for all teams in the National League through time. The trend shows greater equalization of teams in the history of baseball, a consequence of increasing general excellence of play.

by combination of genetic gift, maniacal dedication, and rigorous training, push their bodies to perform as close to the right wall as human achievement will allow.

Earlier I discussed the major phenomenon in sports that must be signaling approach to the right wall—a flattening out of improvement (measured by record breaking) as sports mature, promise ever greater rewards, become accessible to all, and optimize methods of training (see pp. 92–97). This flattening out must represent the approach of the best to the right wall. The longer a sport has endured with stable rules and maximal access, the closer the best should stand to the right wall, and the less we should therefore expect any sudden and massive breaking of records. When George Plimpton, several years ago, wrote about a great pitching prospect who could throw 140 miles per hour, all serious fans recognized this essay in "straight" reporting as a spoof, though many less knowledgeable folks were fooled. From Walter Johnson in the 1920s to Nolan Ryan today, the best fastball pitchers have tried to throw at maximal speed, and no one has consistently broken 100 mph. In fact, Johnson was probably as fast as Ryan. Thus, we can assume that these men stand near the right wall of what a human arm can do. Barring some unexpected invention in technique, no one is going to descend from some baseball Valhalla and start throwing 40 percent again as fast—not after a century of trying among the very best.

These approaches to the right wall can easily be discerned in sports that keep absolute records measured as times and distances. As previously discussed, record times for the marathon, or virtually any other timed event with stable rules and no major innovations, drop steadily—in the decelerating pattern of initial rapidity, followed by later plateauing as the best draw near to the right wall. But this pattern is masked in baseball, because most records measure one activity relative to another, and not against an absolute standard of time or distance. Batting records mark what a hitter does against pitchers. A mean league batting average of 0.260 is not an absolute measure of anything, but a general rate of success for hitters versus pitchers. Therefore a fall or rise in mean batting average does not imply that hitters are becoming absolutely worse or better, but only that their performance relative to pitchers has changed.

Thus we have been fooled in reading baseball records. We note that the mean batting average has never strayed much from 0.260, and we therefore wrongly assume that batting skills have remained in a century-long rut. We note that 0.400 hitting has disappeared, and we falsely assume that great hitting has gone belly-up. But when we recognize these averages as relative records, and acknowledge that baseball professionals, like all other premier athletes, must be improving with time, a different (and almost surely correct) picture emerges (see Figure 19)—one that acknowledges batting averages as components in a full house of variation with a bell-curve distribution and that, as an incidental consequence of no mean importance, allows us finally to visualize why the extinction of 0.400 hitting must be measuring improvement of play as marked by shrinking variation.

Early in the history of baseball (top part of Figure 19), average play stood far from the right wall of human limits. Both hitters and pitchers performed considerably below modern standards, but the balance between them did not differ from today's—and we measure this unchanging balance as 0.260 hitting. Thus, in these early days, the mean batting average of 0.260 fell well below the right wall, and variation spread out widely on both sides—at the lower end, because the looser and less accomplished system did provide jobs to good fielders who couldn't hit; and

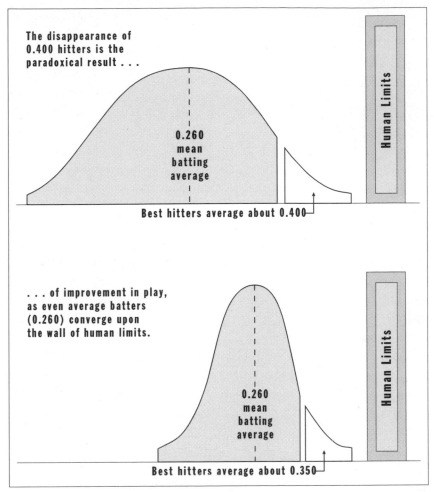

The disappearance of
0.400 hitters is the
paradoxical result . . .

0.260
mean
batting
average

Best hitters average about 0.400

Human Limits

. . . of improvement in play,
as even average batters
(0.260) converge upon
the wall of human limits.

0.260
mean
batting
average

Best hitters average about 0.350

Human Limits

FIGURE 19

Four hundred hitting disappears as play improves and the entire bell curve moves
closer to the right wall of human limits while variation declines. Upper chart: early
twentieth-century baseball. Lower chart: current baseball.

at the upper end, because so much space existed between the average and
the right wall.

A few men of extraordinary talent and dedication always push their
skills to the very limit of human accomplishment and reside near the right
wall. In baseball's early days, these men stood so far above the mean that
we measured their superior performance as 0.400 batting.

Consider what has happened to modern baseball (lower part of Figure 19). General play has improved significantly in all aspects of the game. But the balance between hitting and pitching has not altered. (I showed on pp. 101–105 that the standardbearers of baseball have frequently fiddled with the rules in order to maintain this balance.) The mean batting average has therefore remained constant, but this stable number represents markedly superior performance today (in both hitting *and* pitching). Therefore, this unchanged average must now reside much closer to the right wall. Meanwhile, and inevitably, variation in the entire system has shriveled symmetrically on both sides—at the lower end, because improvement of play now debars employment to men who field well but cannot hit; and at the upper end, for the simple reason that much less room now exists between the upwardly mobile mean and the unchanging right wall. The top hitters, trapped at the upper bound of the right wall, must now lie closer to the mean than did their counterparts of yore.

The best hitters of today can't be worse than 0.400 hitters of the past. In fact, the modern stars may have improved slightly and may now stand an inch or two closer to the right wall. But the average player has moved several feet closer to the right wall—and the distance between ordinary (maintained at 0.260) and best has decreased, thereby erasing batting averages as high as 0.400. Ironically, therefore, the disappearance of 0.400 hitting marks the general improvement of play, not a decline in anything.

Our confidence in this explanation will increase if supporting data can be provided with statistics for other aspects of play through time. I have compiled similar records for the other two major facets of baseball—fielding and pitching. Both support the key predictions of a model that posits increasing excellence of play with decreasing variation when the best can no longer take such numerical advantage of the poorer quality in average performance.

Most batting and pitching records are relative, but the primary measure of good fielding is absolute (or at least effectively so). A fielding average is you against the ball, and I don't think that grounders or fly balls have improved through time (though the hitters have). I suspect that modern fielders are trying to accomplish the same tasks, at about the same level of difficulty, as their older counterparts. Fielding averages (the percent of errorless chances) should therefore provide an absolute measure of changing excellence in play. If baseball has improved, we should note a decelerating rise in fielding averages through time. (I do recognize that some improvement might be attributed to changing conditions, rather than absolutely improving play, just as some running records may fall because modern tracks are better raked and pitched. Older infields were, apparently, lumpier and bumpier than the productions of good ground crews today—so some of the poorer fielding of early days may have resulted from lousy fields rather than lousy fielders. I also recognize that rising averages must be tied in large part to great improvement in the design of gloves— but better equipment represents a major theme of history, and one of the legitimate reasons underlying my claim for general improvement in play.)

Following the procedure of my first compilation on batting averages, I computed both the league fielding average for all regular players and the mean score of the five best for each year since the beginning of major league play in 1876. Figure 20, showing decadal averages for the National League through time, confirms the predictions in a striking manner. Not only does improvement decelerate strongly with time, but the decrease is continuous and entirely unreversed, even for the tiny increments of the last few decades, as averages reach a plateau so near the right wall.

For the first half of baseball (the fifty-five years from 1876 to 1930),

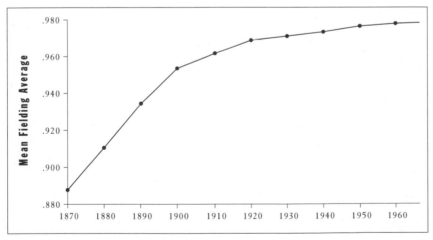

FIGURE 20

Unreversed, but constantly slowing, improvement in mean fielding average through the history of baseball.

decadal fielding averages rose from 0.9622 to 0.9925 for the best players, for a total gain of 0.0303; and from 0.8872 to 0.9685 for average performers, for a total gain of 0.0813. (For a good sense of total improvement, note that the average player of the 1920s did a tiny bit better than the very best fielders of the 1870s.) For baseball's second half (the fifty years from 1931 to 1980), the increase slowed substantially, but never stopped. Decadal averages for the best players rose from 0.9940 to 0.9968, for a small total gain of 0.0028—or less than 10 percent of the recorded rise of baseball's first half. Over the same fifty years, values for league averages rose from 0.971 for the 1930s to 0.9774 for the 1970s, a total gain of 0.0064—again less than 10 percent of the improvement recorded during the same number of years during baseball's first half.

These data continue to excite me. As stated before, I have spent a professional lifetime compiling statistical data of this sort for the growth of organisms and the evolution of lineages. I have a sense of the patterns expected from such data, and have learned to pay special attention to noise and inevitable departures from expectations. I am just not used to the exceptionless data produced over and over again by the history of baseball. I would have thought that any human institution must be more sensitive than natural systems to the vagaries of accident and history, and that baseball would therefore yield more exceptions and a fuzzier signal (if any at all). And yet, here again—as with the decline of standard deviations in batting averages (see page 106)—I find absolute regularity of change, even when the total accumulation is so small that one would expect some exceptions just from the inevitable statistical errors of life and computation. Again, I get the eerie feeling that I must be calculating something quite general about the nature of systems, and not just compiling the individualized numbers of a particular and idiosyncratic institution (yes, I know, it's just a feeling, not a proof). Baseball is a truly remarkable system for statisticians, manifesting two properties devoutly to be wished, but not often encountered, in actual data: an institution that has worked by the same rules for a century, and has compiled complete data (nothing major missing) on all measurable aspects of its history.

For example, as decadal averages for the five best reach their plateau in baseball's second half, improvement slows markedly, but never reverses—the total rise of only 0.0028 occurs in a steady climb by tiny increments: 0.9940, 0.9953, 0.9958, and 0.9968. Lest one consider these gains too small to be anything but accidental, the first achievements of individual yearly values also show the same pattern. Who would have thought that the rise from 0.990 to 0.991 to 0.992, and so on, could mean anything at all? An increment of one in the third decimal place can't possibly be measuring anything significant about actual play. And yet 0.990 is first reached in 1907, 0.991 in 1909, 0.992 in 1914, 0.993 in 1915, 0.994 in 1922, 0.995 in 1930. Then, thank goodness, I find one tiny break in pattern (for I was beginning to think that baseball's God had decided to mock me; the natural world is supposed to contain exceptions). The first value of 0.996 occurs in 1948, but the sharp fielders of 1946 got to 0.997 first! Then we are back on track and do not reach 0.998 until 1972.

This remarkable regularity can occur only because, as my hypothesis requires in its major contention, variation declines so powerfully through time and becomes so restricted in later years. (With such limited variation from year to year, any general signal, however weak, should be more easily detected.) For example, yearly values during the 1930s range only from 0.992 to 0.995 for best scores, and from 0.968 to 0.973 for average scores. By contrast, during baseball's first full decade of the 1880s, the yearly best ranged from 0.966 to 0.981, and the average from 0.891 to 0.927.

This regularity may be affirmed with parallel data for the American League (shown with the National League in Table 3). Again, we find unreversed decline, though this time with one exception as American League values fall slightly during the 1970s—and I have no idea why (if one can properly even ask such a question for such a minuscule effect). Note the remarkable similarity between the leagues in rates of improvement across decades. We are not, of course, observing two independent systems, for styles of play do alter roughly in parallel as both leagues form a single institution (with some minor exceptions, as the National League's blessed refusal to adopt the designated hitter rule indicates in our times). But nearly identical behavior in two cases does show that we are probably picking up a true signal and not a statistical accident.

Data on fielding averages are particularly well suited to illustrate the focal concept of right walls—the key notion behind my second explanation for viewing the disappearance of 0.400 hitting as a sign of general improvement in play. Fielding averages have an absolute, natural, and logical

TABLE 3

DECADAL FIELDING AVERAGES FOR FIVE BEST PLAYERS AND FOR ALL
PLAYERS IN MAJOR LEAGUE BASEBALL

| | NATIONAL LEAGUE | | AMERICAN LEAGUE | |
	All Players	Five Best	All Players	Five Best
1870s	.8872	.9622		
1880s	.9103	.9740		
1890s	.9347	.9852		
1900s	.9540	.9874	.9543	.9868
1910s	.9626	.9912	.9606	.9899
1920s	.9685	.9925	.9681	.9940
1930s	.9711	.9940	.9704	.9946
1940s	.9736	.9953	.9740	.9946
1950s	.9763	.9955	.9772	.9960
1960s	.9765	.9958	.9781	.9968
1970s	.9774	.9968	.9776	.9967

right wall of 1.000—for 1.000 represents errorless play, and you cannot make a negative number of errors! Today's best fielders are standing with toes already grazing the right wall—0.998 is about an error per year, and nobody can be absolutely perfect. (Outfielders, pitchers, and catchers occasionally turn in seasons of 1.000 fielding, but only one infielder has ever done so for a full season's regular play—Steve Garvey at first base in 1984.)

If you doubted my explanation for shrinking variation at the upper end of the bell curve for batting averages—that as the mean moves toward the right wall, variation scrunches up into an ever smaller available space, and must therefore decrease—you will surely grant me the argument for fielding averages so close to an absolute wall. Even the 1870s didn't provide much space, but fielders had a bit of breathing room for improvement between their first decadal best of 0.962 and the wall. And improve they did, and steadily. But now, with the five best averaging 0.9968, there just isn't any more space, barring the construction of truly errorless robotic fielding machines.

As the mean moves toward the wall, variation must decrease. For absolute measures of fielding, high numbers persist and low values get axed. But for relative measures of hitting, the wall itself bears no number. The advancing mean retains the same value (as a balance between hitting and pitching), while both hitting and pitching move in lockstep toward their right walls of human limitation. Thus, 0.400 hitting disappears as the league mean of 0.260 marches steadily toward the wall. But the 0.400 hitters of yore are alive and well, probably more numerous than ever, and standing where they always have resided—just inches from the right wall. But their current best does not measure 0.400 anymore, because everyone else has improved so much, raising average play to a level where an unchanged (or even slightly improved) best can no longer soar so far above the norm.

The best hitters of early baseball could compile 0.400 averages by taking advantage of a standard in average play much lower than today's premier batters encounter. Wade Boggs would hit 0.400 every year against the pitching and fielding of the 1890s, while Wee Willie Keeler would be lucky to crack 0.320 today. Since pitching and batting both feature relative records, and presumably exist in effective balance throughout the history of baseball, we should be able to detect similar phenomena in the statistics of pitching through time. The best pitchers of the past, legendary figures like Christy Mathewson, Cy Young, Walter Johnson, Three Finger Brown, and Grover Cleveland Alexander, should be no better than their modern counterparts Sandy Koufax, Bob Gibson, Tom Seaver, and Nolan Ryan. But the old pitchers, standing next to their own right wall and facing much poorer average batting, should have racked up numbers that modern hurlers just can't equal.

The fascinating and well-known history of minimal earned run averages provides our best illustration of symmetry between batting and pitching—another indication that these statistics record the general behavior of systems, not just a peculiarity of batting in baseball. As the best batters sacrificed their 0.400 averages because variation declined while average play improved, the best pitchers lost their earned run averages below 1.50 because ordinary hitters became too good.

The list of the hundred best seasonal ERAs shows a remarkable imbalance. More than 90 percent of the entries were achieved before 1920. Since then, only nine pitchers have obtained an earned run average in the top one hundred (and remember that the number of pitchers, hence the number of opportunities, has expanded dramatically, first with the introduction of the American League and later with expansion from an original eight to our current roster of fourteen teams per league). Moreover, of these nine modern values, seven rank in the lower half. If we consider the modern achievements, from the bottom up, we get a good sense of the obstacles that must face our superb contemporary pitchers.

Tied at number 100 are Sandy Koufax (1.74 in 1964) and Ron Guidry (1.74 in 1978). Koufax was, well, Koufax—by general agreement the greatest of modern pitchers, perhaps of all pitchers anywhere, anytime (he also holds the ninety-seventh spot at 1.73 for 1966). Guidry, a wonderful Yankee pitcher for a few years, compiled a stellar season in 1978 (with an unmatched combination of total victories and winning percentage of 25–3, for 0.893), and then threw his arm out. Nolan Ryan occupies eighty-seventh place at 1.69 for 1981. And Ryan was, well, Ryan. Nothing else need be said. Carl Hubbell, perhaps the premier pitcher of the 1930s (Lefty Grove was no slouch, either) turned in 1.66 in 1933 for seventy-sixth place and the only entry for his high-hitting decade. Dean Chance, a strictly okay pitcher of the last generation, posted an anomalous 1.65 for seventy-first place in 1964—and I can't figure this one at all. Spud Chandler holds sixty-sixth place at 1.64 for 1943—a fine (if not fabulous) pitcher during the war years, when all decent hitters were blasting away at Germany or Japan instead. Luis Tiant, a damned fine pitcher but not among the greatest, holds sixtieth place at 1.60 for 1968—and I'll return to him in a moment. Dwight Gooden had a fabulous sophomore season in 1985, with a 1.53 ERA that puts him in forty-second place as one of only two modern pitchers in the first half-hundred. He then fell victim to what the newspapers politely call "substance abuse."

We then come to what may be the finest record in modern sports—

Bob Gibson's truly incredible 1.12 ERA of 1968, for fourth place, surrounded by forty old-timers before we meet Doc Gooden at number forty-two. Gibson's only superiors are Tim Keefe with 0.86 in 1880, Dutch Leonard at 0.96 for 1914, and Three Finger Brown at 1.04 for 1906. How could Gibson compile such a record—the only post-1920 value below 1.50, and way, way below at that—in our modern era of greatly improved average hitting?

I don't want to take a thing away from Bob Gibson, who absolutely terrified me in the 1967 World Series, when he almost single-handedly beat the Red Sox by winning three games and casting a pall of inevitability over the whole proceedings. But, in slight mitigation, 1968 was a really funny year, as mentioned previously (see page 104). For some set of reasons that no one understands, pitching took a dramatic upper hand that year, capping a trend of several years' duration. (As explained before, the rule-makers then restored the usual order by lowering the pitching mound and decreasing the strike zone; batting averages and ERAs rose appropriately in the 1969 season and have remained in balance ever since.) The 1968 season didn't just belong to Gibson; in that year, low ERAs sprouted like dandelions in my garden. In most years of modern baseball, no pitcher in either league has posted an ERA lower than 2.00. Uniquely in 1968, all five leading American League pitchers bettered this mark, as Yastrzemski won the batting title with a paltry 0.301 (Tiant at 1.60, McDowell at 1.81, McNally at 1.95, McLain at 1.96—a banner year for Scotland—and John at 1.98. As I said, Tiant was a terrific pitcher and great fun to watch, but not one of the game's greatest. If he could post 1.60 for 1968, baseball was really out of whack that year.) So Gibson certainly took maximal advantage of a weird year, but let's not take anything away from him. No one, no matter how good, had any statistical right to post a value so much better than anything achieved for sixty years, especially when general improvement in play should have made such low ERAs effectively unobtainable. Gibson had one helluva year!

In quick summary of a long and detailed argument, symmetrically shrinking variation in batting averages must record general improvement of play (including hitting, of course) for two reasons—the first (expressed in terms of the history of institutions) because systems manned by best performers in competition, and working under the same rules through time, slowly discover optimal procedures and reduce their variation as all personnel learn and master the best ways; the second (expressed in terms of performers and human limits) because the mean moves toward the right wall, thus leaving less space for the spread of variation. Hitting 0.400 is not a *thing,* but the right tail of the full house for variation in batting averages. As variation shrinks because general play improves, 0.400 hitting disappears as a consequence of increasing excellence in play.

·11·

A Philosophical Conclusion

Some people regard this explanation as a sad story. One can scarcely decry a general improvement in play, but the increasing standardization thus engendered does seem to remove much of the fun and drama from sports. The "play" in play diminishes as activities become ever more "scientific" in the pejorative sense of operating like optimized clockwork. Perhaps no giants inhabited the earth during baseball's early days, but the best then soared so far above the norm that their numbers seemed truly heroic and otherworldly, while our current champions cannot rise nearly so far above the vastly improved average.

But I suggest that we should rejoice in the shrinkage of variation and consequent elimination of 0.400 hitting. Yes, excellence in play does imply increasing precision and standardization, but what complaint can we lodge against repeated maximal beauty? I have now been a fan for fifty years. I have seen hundreds of perfectly executed double plays and brilliant pegs from outfield to home (that may or may not have beaten the runner charging from third)—the kind of beautifully orchestrated precision that probably occurred only rarely in baseball's early years. I do not thrill any less with each repetition. The pinnacle of excellence is so rare, its productions so exquisite. Did we ever get bored with Caruso or Pavarotti in their prime? I would much rather have my expectation of excellence affirmed when I go to the ballpark or the opera house than to take potluck and hope for a rare glimpse of glory in a sea of mediocrity.

Moreover, the rise in general excellence and consequent shrinkage of variation does not remove the possibility of transcendence. In fact, I would argue that transcendence becomes all the more intriguing and exciting for the smaller space now allocated to such a possibility, and for the consequently greater struggle that must attend the achievement. When the norm stood miles from the right wall, records could be broken with relative ease. But when the average player can almost touch the wall, then transcendence of the mean marks a true outer limit for conceivable human achievement. (Again, I would make an analogy to musical performance. Do we not rejoice when every string in a symphony orchestra plays with exquisite beauty and consummate professionalism? And do we not thrill all the more when, in this context of superb general performance, a great soloist does something so special that only angels in heaven could have contemplated the possibility?) I would carry the argument even further and point out that a norm near the right wall pushes the very best to seek levels of greater accomplishment that otherwise might never have been conceptualized. I will speak in the final chapter about the heroic efforts, often with attendant accident and loss of life, that such "pushing of the envelope" imposes in the almost holy mania that infects the greatest performers in the circus arts and other dangerous activities. Call it foolish (and swear up and down that you would never so act yourself), but acknowledge that human greatness often forms a strange partnership with human obsession, and that the mix sometimes spells glory—or death.

The possibility of transcendence can never die, because this pinnacle of admiration in sports can be reached by several attainable routes. First of all, a kind of democracy infests individual games. When we go to the ballpark, we never know what we will witness. At any time, even the worst team may execute a thrilling play with awesome perfection. The event may occur only once a year (or much less often) on average, but the day of your attendance may feature a triple play, a steal of home, a rip-roaring, bench-clearing brawl (yes, as *Homo ludens* and *Homo stupidus,* we also root for this sort of rare nonsense from the underside of our complete lives), or an inside-the-park homer, with the runner just slipping under the catcher's tag. You never know.

The enormous variability of individual performance guarantees that even a mediocre player can, for one day of glory, accomplish something never done before, or even dreamed of in baseball's philosophy. Harvey Haddix was a fine pitcher, but not the greatest. Yet one day he hurled twelve innings of perfect ball—and then lost the game in the thirteenth (as the opposing pitcher had shut out Haddix's side for the first twelve innings). Bobby Thomson was a better-than-average outfielder for the New York Giants, but one day in 1951 he hit a home run, perfectly ordinary by the physics of distance, but meaningful beyond measure in baseball's enclosed system, because this single blow won the pennant for the Giants against their archrivals, the Brooklyn Dodgers, in the last inning of the last game of a play-off series culminating the greatest comeback in the history of baseball (the Giants had trailed the Dodgers by thirteen and a half games in August, and had entered this last inning with an apparently insurmountable three-run deficit). I was a ten-year-old Giants fan watching the game on our family's first television set, and I have never been so thrilled in all my life (except for one other time).

Don Larsen was a truly mediocre pitcher for the Yankees, but he achieved baseball's definition of perfection when it mattered most: twenty-seven Dodgers up, twenty-seven Bums down on October 8, 1956, for a perfect game in the World Series (no one before or since had ever thrown a no-hitter of any kind in a World Series game). I was a fifteen-year-old Yankees fan (many New Yorkers rooted for two teams, one local club in each league), trying to persuade my French teacher to let us listen to the game's end on the radio. I have never been so thrilled in all my life (except for one other time).

When we move to the statistics of seasonal or lifetime performance, this kind of democracy vanishes, and only the truly great can achieve transcendence. But some humans can push themselves, by an alchemy of inborn skill, happy fortuity, and maniacal dedication, to performances that just shouldn't happen—and we revel when such a man reaches farther and actually touches the right wall. Bob Gibson had no business compiling an ERA of 1.12 in 1968. And I can show you with copious statistics that Joe DiMaggio should never have hit in fifty-six straight games in 1941 (see Gould, 1988). I delayed writing the last paragraph of this chapter for several days because I couldn't bear not to share vicariously in a great moment of transcendence. So I am sitting at this old typewriter on September 6, 1995, as Cal Ripken plays his 2131st consecutive game, eclipsing the "unbreakable" record of the Iron Horse, Lou Gehrig.

No records lie beyond fracture (unless rules or practices have changed to make an old achievement unattainable in modern performance). Perhaps I have exaggerated by discussing the "extinction" of 0.400 hitting in this section. (I am a paleontologist and hate to avoid one of the favorite words in my trade.) But I meant extinction in the literal sense of snuffing out a candle that might be lit again, not in the evolutionary and ecological meaning of species death where, by an accurate motto of our times, extinction is truly forever.

I am not arguing that no one will ever hit 0.400 again. I do say that such a mark has become a consummate rarity, achieved perhaps once in a century like a hundred-year flood, and not the common pinnacle of baseball's early years. The fifty-year drought since Ted Williams supports this view, and I think that this part has identified the reason by reconceptualizing 0.400 hitting as the right tail in a shrinking bell curve of batting averages with a stable mean—all as a necessary and predictable consequence of general improvement in play. But someday, someone will hit 0.400 again—though this time the achievement will be so much more difficult than ever before and therefore so much more worthy of honor. When the idiots on both sides in the great pissing contest of 1994 (otherwise known as a labor dispute) aborted the season and canceled the World Series, Tony Gwynn was batting 0.392 and on the rise. I believe that he would have succeeded had the season unfolded as history and propriety demanded. Someday, someone will join Ted Williams and touch the right wall against higher odds than ever before. Every season brings this possibility. Every season features the promise of transcendence.

Part Four

THE MODAL BACTER:
WHY PROGRESS DOES NOT
RULE THE HISTORY
OF LIFE

·12·

The Bare Bones of Natural Selection

I quote verbatim from a discussion held in 1959:

> HUXLEY: I once tried to define evolution in an overall way somewhat along these lines: a one way process, irreversible in time, producing apparent novelties and greater variety and leading to higher degrees of organization.
>
> DARWIN: What is "higher"?
>
> HUXLEY: More differentiated, more complex, but at the same time more integrated.
>
> DARWIN: But parasites are also produced.
>
> HUXLEY: I mean a higher degree of organization in general, as shown by the upper level attained.

Charles Darwin died in 1882, Thomas Henry Huxley in 1895—so, unless I am reporting a seance, something strange is going on here. The date of 1959 might give a hint for aficionados, for Charles Darwin published the *Origin of Species* in 1859, and the interval of exactly one hundred years smells of a centennial celebration. Huxley, in fact, is Thomas Henry's grandson Julian, an eminent biologist and statesman in his own right, while Darwin is Charles's grandson, also Charles, and also a scientist and social thinker. The two grandsons held their dialogue at the largest centennial celebration for Charles Darwin's *Origin of Species,* held at the University of Chicago in 1959 and published in 1960 as an influential three-volume work edited by Sol Tax.

Not only did the Darwin and Huxley clans maintain a genealogical tradition for evolutionary studies, but also, and more curiously as we shall see, the errors and insights of modern Chicago's Darwin and Huxley closely parallel the attitudes of their blood ancestors. Julian makes the same errors as Thomas Henry; Charles offers some of the same correctives as the elder Charles. Both are confused on the notion of progress. Darwin asks a good question about parasites—and so did his grandfather. Julian Huxley gives a muddled answer that contains the germ of resolution within the standard, central confusion.

The problem that spawns this confusion within the Darwinian tradition may be simply stated as a paradox. The basic theory of natural selection offers no statement about general progress, and supplies no mechanism whereby overall advance might be expected. Yet both Western culture and the undeniable facts of a fossil record that started with bacteria alone, and has now produced exalted us, cry out in unison for a rationale that will place progress into the center of evolutionary theory.

Charles Darwin reveled in the radical nature of his biological philosophy. His early and private notebooks practically shout for joy at the outrageous character of his valid conjectures. He writes to himself, for example, that our feelings of reverence for God arise from some feature of our neurological organization. Only our arrogance, he continues, makes us so reluctant to ascribe our thoughts to a material substrate:

> Love of the deity [an] effect of organization, oh you materialist! . . . Why is thought being a secretion of brain,

more wonderful than gravity a property of matter? It is
our arrogance, our admiration of ourselves.

Darwin toned down his exultation as he grew older and presented his
work for public appraisal, but he never abandoned his radical perspective—and we have therefore, as discussed in Part One, never been able or
willing to complete his revolution in Freud's sense by owning the true implications of Darwinism for the dethronement of human arrogance. None
of Darwin's *outré* ideas could have been more unacceptable than his denial of progress as a predictable outcome of the mechanisms of evolutionary change. Most other nineteenth-century evolutionists, including
Lamarck, presented much more congenial theories that did include predictable progress as a central ingredient. In fact, *evolution* entered our language as the favored word for what Darwin had called "descent with
modification" because most Victorian thinkers equated such biological
change with progress—and the word *evolution,* propelled into biology by
the advocacy of Herbert Spencer, meant progress (literally "unfolding")
in the English vernacular. Darwin initially resisted the word because his
theory embodied no notion of general advance as a predictable consequence of any mechanism of change. *Evolution* never appears in the first
edition of the *Origin of Species,* and Darwin first used the word in *The Descent of Man* in 1871. He never liked *evolution,* and only acquiesced because
Spencer's term had gained general currency.

Darwin was not shy in advertising his nonprogressivism. He jotted a
note in the margins of a major book that did advocate progress in the history of life: "Never say higher or lower." He wrote the following line in a
letter (December 4, 1872) to the paleontologist Alpheus Hyatt, who had
proposed an evolutionary theory based on intrinsic progress (I now inhabit
Hyatt's old office, so the connection has a special meaning for me): "After
long reflection, I cannot avoid the conviction that no innate tendency to
progressive development exists."

Darwin's denial of progress arises for a special and technical reason
within his theory, and not merely from a general philosophical preference.
In a famous anecdote, T. H. Huxley, upon first learning the content of Darwin's theory of natural selection, proclaimed himself "extremely stupid"
not to have figured out this principle by himself. Unlike other celebrated

(and truly arcane) ideas in the history of science, natural selection is a remarkably simple notion—basically three undeniable facts followed by an obvious, almost syllogistic conclusion. (I speak of simplicity only for the "bare bones" of natural selection as a mechanism; the inferences and implications that flow from the operation of selection can be quite subtle and complex.)

Darwin devotes the beginning chapters of the *Origin of Species* to validating the three facts:

1. All organisms tend to produce more offspring than can possibly survive (Darwin's generation gave this principle the lovely name of "super-fecundity").

2. Offspring vary among themselves, and are not carbon copies of an immutable type.

3. At least some of this variation is passed down by inheritance to future generations. (Darwin did not know the mechanism of heredity, for Mendel's principles did not gain acceptance until early in our century. However, this third fact requires no knowledge of how heredity works, but only an acknowledgment that heredity exists. And mere existence is undeniable folk wisdom. We know that black folks have black kids; white folks, white kids; tall parents tend to have tall children; and so on.)

The principle of natural selection then emerges as a necessary inference from these facts:

4. If many offspring must die (for not all can be accommodated in nature's limited ecology), and individuals in all species vary among themselves, then on average (as a statistical statement, and not in every case), survivors will tend to be those individuals with variations that are fortuitously best suited to changing local environments. Since heredity exists, the offspring of survivors will tend to resemble their successful parents. The accumulation of these favorable variants through time will produce evolutionary change.

If this presentation seems overly abstract, consider a potential example (something of a simplistic caricature, to be sure, but not bad as a representation of the central features of Darwin's argument): an earlier Siberia is nicely temperate, and a population of minimally hairy elephants dwells there in excellent adaptation. As the earth enters a glacial age, and ice begins to build up to the north, climates become colder and possession of more than the usual amount of hair becomes a decided advantage. On average, the hairier elephants will be more successful and therefore leave more surviving offspring. (On average, that is, and not every time—the hairiest elephant in the population can still slip into a crevasse and die.) Since hairiness is inherited, the next generation will contain more elephants with increased hair (for the hairiest of the last generation enjoyed greater success in reproduction). Continue this process for a large number of generations, and eventually Siberia will house a population of woolly mammoths—the evolutionary descendants of the original elephants.

Fine, in outline. But note what this scenario leaves out (that almost all popular views of evolution include as a defining feature). Natural selection talks only about "adaptation to changing local environments"; the scenario includes no statement whatever about progress—nor could any such claim be advanced from the principle of natural selection. The woolly mammoth is not a cosmically better or generally superior elephant. Its only "improvement" is entirely local; the woolly mammoth is better in cold climates (but its minimally hairy ancestor remains superior in warmer climates). Natural selection can only produce adaptation to immediately surrounding (and changing) environments.

No feature of such local adaptation should yield any expectation of general progress (however such a vague term be defined). Local adaptation may as well lead to anatomical simplification as to greater complexity. As an adult, the famous parasite *Sacculina,* a barnacle by ancestry, looks like a formless bag of reproductive tissue attached to the underbelly of its crab host (with "roots" of equally formless tissue anchored within the body of the crab itself)—a devilish device to be sure (at least by our aesthetic standards), but surely less anatomically complex than the barnacle on the bottom of your boat, waving its legs through the water in search of food.

If a sequence of local environments could elicit progressive advance through time, then some expectation of progress might be drawn from natural selection. But no such argument seems possible. The sequence of local environments in any one place should be effectively random through geological time—the seas come in and the seas go out, the weather gets colder, then hotter, etc. If organisms are tracking local environments by natural selection, then their evolutionary history should be effectively random as well.

These arguments led Darwin to his denial of progress as a consequence of the "bare bones mechanics" of natural selection—for this process yields only local adaptation, often exquisite to be sure, but not universally advancing. The mammoth is every bit as good as an elephant—and vice versa. Do you prefer a marlin for its excellent spike; a flounder for its superb camouflage; an anglerfish for its peculiar "lure" evolved at the end of its own dorsal fin ray; a seahorse for its wondrous shape, so well adapted for bobbing around its habitat? Could any of these fishes be judged "better" or "more progressive" than any other? The question makes no sense. Natural selection can forge only local adaptation—wondrously intricate in some cases, but always local and not a step in a series of general progress or complexification.

Darwin reveled in this unusual feature of his theory—this mechanism for immediate fit alone, with no rationale for increments of general progress or complexification. So far, so good; so logical, so clear. I should end my discussion of Darwin right here, extolling him as a consistent intellectual radical whose vision of a history of life devoid of predictable progress proved too much for his Western compatriots to accept.

Simple, and heroic for Darwin, but quite untrue—for real history (and biography) tends to be much messier. Actual lives, especially for brilliantly complex men like Darwin, abound in pieces that don't quite mesh, or that truly contradict. Darwin was intellectually radical; but he was also politically liberal, a defender of mild social reform and a passionate opponent of slavery; and decidedly conservative in lifestyle—a wealthy country squire himself, reared in the same background, and with no desire to change the amenities of his comfortable existence.

Moreover, Darwin enjoyed this comfort in a society that, more than any other in human history, had enshrined progress as the fundamental doctrine of its meaning and being—Victorian Britain at the height of industrial and colonial expansion. How could a patrician Englishman, at the very apex of his nation's thundering success, abjure the principle that embodied this triumph? And yet, natural selection could produce only local adaptation, not general progress. How could these contradictory needs—the intellectual and the social—be reconciled?

These conflicting loyalties achieve their sharpest expression in a remarkable sentence that Darwin placed in a most conspicuous position—on the last page of the *Origin of Species,* just before the famous concluding paragraph about "grandeur in this view of life."

> As natural selection works solely by and for the good of
> each being, all corporeal and mental endowments will
> tend to progress towards perfection.

Note the starkness of the claim. Darwin says "all" endowments—including *all* attributes of *mind,* as well as all features of bodies. How, after proclaiming with such panache (as quoted earlier) that natural selection undermines the old dogma of progress, could Darwin write such a sentence?

Darwin's apparent contradictions on the subject of progress have sparked a large literature among historians of science. Entire books have been dedicated to the subject (see Richards, 1992). Most efforts have been devoted to constructing forced and arcane rationales that would render all of Darwin's statements consistent. But I would take another view, based on Emerson's famous dictum that "a foolish consistency is the hobgoblin of little minds," or upon Walt Whitman's wonderful lines in his "Song of Myself":

> *Do I contradict myself?*
> *Very well then, I contradict myself,*
> *(I am large, I contain multitudes).*

I believe that Darwin's views contain an unresolved inconsistency. Darwin, the intellectual radical, knew what his own theory entailed and implied; but Darwin, the social conservative, could not undermine the defining principle of a culture (at a key moment of history) to which he felt such loyalty, and in which he dwelt with such comfort.

Darwin did, of course, supply an argument to bridge the two starkly contradictory claims—that the mechanics of natural selection produces only local adaptation, not general progress; and that all mental and corporeal endowments advance toward perfection in the history of life. He could not, after all, leave such a gaping logical hole in his *oeuvre*. Darwin tried to plug the hole by adding a set of statements about ecology to the "bare bones mechanics" that could not, by itself, validate progress.

Darwin began by drawing a distinction between two kinds of "struggle" in his famous phrases—"struggle for existence" and "survival of the fittest." Struggle may take place directly against other organisms for limited resources (a type of competition called *biotic*), or against the rigors of the physical environment (called *abiotic,* or not involved with other living forms):

> I should premise that I use the term Struggle for Existence in a large and metaphorical sense. . . . Two canine animals in a time of dearth, may be truly said to struggle with each other which shall get food and live. But a plant on the edge of a desert is said to struggle for life against the drought (1859, page 62).

Abiotic competition (the plant at the edge of the desert) cannot yield progress, for physical environments do not change in a persistent direction through time, and local adaptation can produce only a set of backings and forthings as lineages evolve first one way and then the other. But Darwin felt that biotic competition (two canine animals in a time of dearth) might yield progress—for if you are struggling with other members of your own species, rather than against a physical habitat, a more general biomechanical improvement transcending the particulars of any given environment—running faster, enduring longer, thinking better—might be your best option under natural selection. Thus, Darwin continued, if biotic competition is much more important than abiotic competition in the history of life, a general trend to progress might be defended.

But this argument for the prevalence of biotic competition will not suffice; another step is required. If environments are relatively empty—either because defeated forms can migrate somewhere else, or because losers can survive by switching to some other food or space in the same environment—then biomechanically inferior forms can continue to exist, and no ratchet to general progress will exist. But if ecologies are always chock-full of species, and losers have no place to go, then the victors in biotic competition will truly eliminate the vanquished—and the buildup of these successive eliminations might produce a trend to general progress. In fact, Darwin strongly advocated such a concept of nature's plenitude—a notion that he tended to defend with a striking metaphor of the "wedge." Darwin depicts nature as a surface covered with wedges hammered into the ground and filling all space. A new species (depicted as a homeless wedge) can find a dwelling place only by discovering a tiny space between two existing wedges and hammering itself in by forcing another wedge out. In other words, each entry requires an expulsion—and biomechanical improvement might be the general key to successful wedging:

> Nature may be compared to a surface covered with ten thousand sharp wedges . . . representing different species, all packed closely together and all driven in by incessant blows . . . sometimes a wedge . . . driven deeply in forcing out others; with the jar and shock often transmitted far to other wedges in many lines of direction (from an 1856 manuscript, published by Stauffer, 1975).

Darwin then summed up his argument about biotic competition in a persistently full world by writing this and similar passages in the *Origin of Species:*

> The inhabitants of each successive period in the world's history have beaten their predecessors in the race for life, and are, in so far, higher in the scale of nature; and this may account for that vague yet ill-defined sentiment, felt by many paleontologists, that organization on the whole has progressed (1859, page 345).

I do not say that any obvious error pervades the logic of this argument, but we do need to inquire why Darwin bothered, and why the issue seemed important to him. Darwin had just devised an argument against progress—that the "bare bones mechanics" of natural selection yields only local adaptation, not general advance—and he had reveled in the radical character of this claim. Why, then, did he bother to smuggle progress back in through the rear door of a complex and dubious ecological argument about the predominance of biotic competition in a persistently full world? (Darwin surely recognized the shaky character of his necessary premise. He provided no clear rationale for biotic predominance—and Kropotkin and other critics would nail him on this point later. And the fossil record argued strongly against a persistently full world on a crucial issue that caused Darwin no end of trouble. Life's history has been punctuated with several episodes of mass extinction; the largest, at the end of the Permian period, 250 million years ago, wiped out some 95 percent of the species of marine invertebrates. Clearly, habitats could not have been full after such episodes. Therefore, any buildup of progress between mass extinctions should be undone by the next dying. Darwin feared this argument greatly, and could extract himself only by claiming that mass extinctions were artifacts of an imperfect fossil record, an idea that can now be disproved with hard evidence for the triggering of at least one great dying by impact of an extraterrestrial body—the Cretaceous event that wiped out dinosaurs and gave us mammals a chance.)

I have no special insight into Darwin's psyche, but I do feel that his strained and uncomfortable argument for progress arises from a conflict between two of his beings—the intellectual radical and the cultural conservative. The society that he loved, and that brought him such reward, had enshrined progress as its watchword and definition (I think of Herbert Spencer's famous essay "Universal Progress, Its Law and Cause"). Darwin could not bear to fail his own world by denying its central premise. Yet his basic theory required just this opposition. So he forged an escape, and concocted a tenuous resolution by scaffolding a separate argument about ecology onto an edifice that could not support the required proposition by its own unique and different strength. But buildings with scaffolds look messy and incomplete—so why erect such a covering over a lovely structure that stands ever so well all by itself? I know no better illustration of the cultural power that progress holds over us than this story of Darwin's own unresolved intellectual struggle, this tug-of-war between the logic of his theory and the needs of his society. If Darwin could not liberate himself from this deepest presupposition of our shared culture—even after inventing the theory with the key to this conceptual lock—then why should we be doing any better?

Fine. We may identify our assumption that evolution must entail progress as a cultural bias, and we may recognize that no good scientific argument for expecting progress exists, no more so in our own time than in Darwin's day. We may also acknowledge that all standard attempts, including Darwin's own, lie mired in social presupposition for the impetus, logical weakness for the argument, and factual inadequacy for the evidence.

And yet, undeniably (even for such curmudgeons as me), a basic fact of the history of life—*the* basic fact, one might well say—seems to cry out for progress as the central trend and defining feature of life's history. The first fossil evidence of life, from rocks some 3.5 billion years in age, consists only of bacteria, the simplest forms that could be preserved in the geological record. Now we have oak trees, praying mantises, hippopotamuses, and people. How could anyone deny that such a history displays progress above anything else?

But every apparent certainty breeds subsequent doubt. Yes, peccaries, petunias, and poetry. But the earth remains chock-full of bacteria, and insects surely dominate among multicellular animals—with about a million described species versus only four thousand or so for mammals. If progress is so damned obvious, how shall this elusive notion be defined when ants wreck our picnics and bacteria take our lives? This very confusion permeates the fascinating colloquy between the Huxley and Darwin grandchildren, as quoted at the beginning of this chapter. The modern Darwin asks the right questions, just as his grandfather did: How can "higher" be defined in an evolutionary world that produces a parasite for each supposed gain in progress? The modern Huxley gives a confused answer that unknowingly contains the germ of resolution: "I mean a higher degree of organization in general, as shown by the upper level attained." But to grasp the germ and unravel the confusion, we must reconceptualize the entire subject in a fundamental way—the same way that allowed us to resolve the paradox of 0.400 hitting; the same way that forms the subject of this entire book: viewing a history of change as the increase or contraction of variation in an entire system (a "full house"), rather than as a "thing" moving somewhere.

Claims for progress represent a quintessential example of conventional thinking about trends as entities on the move. From life's infinite variety, we extract some "essential" measure like "average complexity" or "most complex creature"—and we then trace the supposed increase of this entity through time (as illustrated in the opening example of this book—see Figure 1). We label this trend to increase as "progress"—and we are locked into the view that such progress must be the defining thrust of the entire evolutionary process.

I shall, for the rest of Part Four, follow the same strategy of all my other examples by trying to view the variation in life's complexity as primary and irreducible. I shall then trace the history of this variation through time. Only in this more adequate way can we acknowledge the obvious fact of "once only bacteria, but now petunias and people as well"—and still understand that no pervasive or predictable thrust toward progress permeates the history of life. We will, in short, learn the deeper reason why Darwin was right when he granted his radical intellect sway over his traditional social values.

·13·

*A Preliminary Example
at Smallest Scale, with
Some Generalities on the
Evolution of Body Size*

In the case of 0.400 hitting, I spoke of a limit or "right wall" of human biomechanical possibility, and I illustrated the decrease in variation of batting averages as the full house of hitters moved toward this upper bound. In this section on complexity in the history of life, I shall present something close to a "mirror image" case—an *increase in total variation by expansion away from a lower limit, or "left wall,"* of simplest conceivable form. The cases may seem quite different at first: improvement in base-ball as decrease in variation by scrunching up against a right wall of max-imal achievement versus increase of variation by spread away from a left wall of minimal complexity, misconstrued as an inevitable, overall march to progress in the history of life.

But a vital and deeper similarity unites the two examples—for both represent the same mode of correction for the same kind of error. In both cases, the error involves false portrayal of a complete system of variation by a single "thing" or entity construed as either the average or the best example within the system. Thus we tried to map the changing status of batting through time by tracing the history of the best conceived as a separable entity (0.400 hitters). Since this "thing" disappeared through time, we naturally assumed that the entire phenomenon—hitting in general—had gotten worse in some way. But proper consideration of the full house—the bell curve of batting averages for *all* regular players—shows that 0.400 hitting (properly viewed as the right tail of this bell curve, and not as a separable "thing") disappeared because variation decreased around a constant mean batting average. I then argued that we must interpret this shrinkage of variation as an indication of general improvement in play through time. In other words, by falsely isolating 0.400 hitting as a thing to be traced by itself, we got the whole story entirely backwards. The partial tale of the "thing" alone seemed to indicate degeneration of hitting; proper consideration of changes in the full variation showed that disappearance of 0.400 hitting represents improvement in general play.

We have traditionally made the same error—and must now make the same correction—in studying apparent trends to increasing complexity, or progress in the history of life. Again, we have abstracted the full and rich complexity of life's variation as a "thing"—by taking either some measure of average complexity in a lineage or, more often, the particular case judged "best" (the most complex, the brainiest)—and we have then traced the history of this "thing" through time. Since our chosen "thing" has increased in complexity through time (once bacteria, then trilobites, now people), how could we possibly deny that progress marks the definition and central driving principle of evolution?

But I shall try to make the same correction in this part by arguing that we must consider the history of life's complexity as a pattern of change for the *full system of variation* through time. Under this properly expanded view, we cannot regard progress as a central thrust and defining trend—for life began with a bacterial mode next to the left wall of minimal complexity; and now, nearly 4 billion years later, life retains the same mode in the same position. The most complex creature may increase in elaboration through time, but this tiny right tail of the full house scarcely qualifies as an essential definition for life as a whole. We cannot confuse a dribble at one end with the richness of an entirety—much as we may cherish this end by virtue of our own peculiar residence.

Before presenting the full argument for all of life, I must first explain why a dribble moving in one direction need not represent the directed thrust of causality within a system—but may actually arise as a consequence of *entirely random movement* among all items within the system. I will then demonstrate, in the next section, that apparent progress in the history of life arises by exactly the same artifact—and that, probably, no average tendency to progress in individual lineages exists at all.

I shall first illustrate the argument as an abstraction—using a classic pedagogical metaphor beloved by teachers of probability. Then I shall provide an intriguing actual case for a lineage of fossils with unusually good and complete data. Since we live in a fractal world of "self-similarity," where local and limited cases may have the same structure as examples at largest scale, I shall then argue that this particular case for the smallest of all fossils—single-celled creatures of the oceanic plankton—presents a structure and explanation identical with an appropriate account for the entire history of life. Since we can approach these largely unknown plankters without the strong biases that becloud our consideration of life's full history, we can best move to the totality by grasping this self-similar example of oceanic unicells.

The overall directionality in certain kinds of random motion—an apparent paradox to many—can best be illustrated by a paradigm known as the "drunkard's walk." A man staggers out of a bar dead drunk. He stands on the sidewalk in front of the bar, with the wall of the bar on one side and the gutter on the other. If he reaches the gutter, he falls down into a stupor and the sequence ends. Let's say that the sidewalk is thirty feet wide, and that our drunkard is staggering at random with an average of five feet in either direction for each stagger. (See Figure 21 for an illustration of this paradigm); for simplicity's sake—since this is an abstract model and not the real world—we will say that the drunkard staggers in a single line only, either toward the wall or toward the gutter. He does not move at right angles along the sidewalk parallel to the wall and gutter.

Where will the drunkard end up if we let him stagger long enough

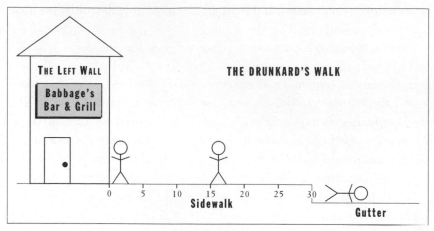

FIGURE 21

The drunkard's walk.

and entirely at random? He will finish in the gutter—absolutely every time, and for the following reason: Each stagger goes in either direction with 50 percent probability. The bar wall at one side is a "reflecting boundary."[8] If the drunkard hits the wall, he just stays there until a subsequent stagger propels him in the other direction. In other words, only one direction of movement remains open for continuous advance—toward the gutter. We can even calculate the average amount of time required to reach the gutter. (Many readers will have recognized this paradigm as just another way of illustrating a preferred result in coin tossing. Falling into the gutter on one unreversed trajectory, after beginning at the wall, has the same probability as flipping six heads in a row [one chance in sixty-four]—five feet with each stagger, to reach the gutter in thirty feet. Start in any other position, and probabilities change accordingly. For example, once the drunkard stands in the middle, fifteen feet from the wall, then three staggers in the same direction [one chance in eight for a single trajectory] put him into the gutter. Each stagger is independent of all others, so pre-

8. In more complex cases involving several entities, the wall might be an "absorbing boundary" that destroys any object hitting it. No matter (so long as enough entities are left to play the game—certainly the case with life's history). The important point is that an entity can't penetrate the wall and continue to move in the wallward direction—whether or not the entity bounces off or gets killed.

vious histories don't count, and you need to know only the initial position to make the calculation.)

I bring up this old example to illustrate but one salient point: In a system of linear motion structurally constrained by a wall at one end, random movement, with no preferred directionality whatever, will inevitably propel the average position away from a starting point at the wall. The drunkard falls into the gutter every time, but his motion includes no trend whatever toward this form of perdition. Similarly, some average or extreme measure of life might move in a particular direction even if no evolutionary advantage, and no inherent trend, favor that pathway.

Turning to a similar example in the history of life, Foraminifera are single-celled protozoans that secrete a skeleton around or within their protoplasm, and are therefore extremely common in the fossil record. (In fact, since they tend to be so abundant—ubiquitous in many marine sediments—they serve as some of the best markers for tracing time and environment in the geological record. Although most of the public never comes in contact with "forams"—as we in the trade call them for short—their study absorbs the lives of a large fraction of professional paleontologists.) Most marine forams live in bottom sediments, and are called *benthic*. A few species float in open waters near the oceanic surface, and are called *planktonic*. These planktonic forams are especially important in dating sediments, and in reconstructing former environments and movements of water masses, during the Cenozoic Era (the past 65 million years, since the extinction of dinosaurs). As a result of their mobility, planktonic species live over large areas of the globe, and are therefore particularly valuable in permitting comparison of sediments from widely separated places (most benthics have much more limited ranges, and consequently less utility).

The basic outline of the evolutionary history of modern planktonic forams has been well known for a long time. They arose in the Cretaceous (the last period of the Mesozoic Era, when dinosaurs dominated terrestrial ecosystems), and they remain vigorously alive today. Their evolution has been interrupted by two episodes of mass extinction, when most species died and only a few survived to continue the lineages: once at the end of the Cretaceous (one of the five great mass dyings in the history of life—this is the event that triggered the death of dinosaurs and almost surely involved the impact of a large extraterrestrial object as a fundamental cause); and again during the largest episode of extinction within the Cenozoic Era. Thus, the evolution of planktonic forams is a drama in three largely independent acts (linked by a few transitions): the Cretaceous for Act One, the earlier Cenozoic (called Paleogene) for Act Two, and the later Cenozoic (called Neogene) for Act Three.

Traditional wisdom, and any textbook, will tell you that each of the three acts follows the same pattern, thus making the entire story so famous in professional circles—for paleontologists crave independent repetition as a test for predictable results (the closest a historical scientist can come to the experimental ideal of replication under identical conditions in a laboratory). The founding lineages for each of the three radiations were small in body size—and size then increased (or so we are told) during each of the three evolutionary diversifications. If an identical result occurs in each of three episodes, then we are probably witnessing an evolutionary generality. In fact, paleontologists treasure this case as our best illustration of the one decent phylogenetic "rule" that the fossil record seems to affirm with copious evidence.

The attempt to establish such "rules," or generalities in evolution during geological time, absorbed much attention in generations past. But the strategy largely fizzled because few proposed "rules" survived the weight of accumulated exceptions in our complex and contingent world of evolutionary change. The one generality that survived, and that seems to hold firm as more evidence accumulates, is known as "Cope's Rule" (after the brilliant and contentious nineteenth-century American vertebrate paleontologist)—the observation that most lineages tend to increase in body size during their evolutionary history. (Like all evolutionary generalizations, "Cope's Rule" identifies a predominant relative frequency, not an absolute statement. Many lineages decrease in size. An increase of size in 70 percent of lineages, when we think that a random world should yield half and half, is more than enough for a "rule" in our trade.)

The evidence, as usually presented, certainly seems to support Cope's Rule for planktonic forams. Figure 22 shows the increase in body size over time for both the largest species and the average of all species during Act One of the Cretaceous Period (figures for Acts Two and Three show the same pattern). I shall not deny this evidence of increase in each act for the largest or average species. But this book is dedicated to providing an en-

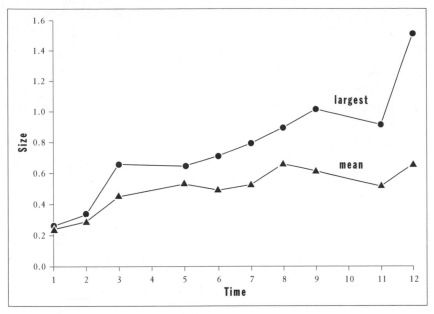

FIGURE 22

Inappropriate use of mean or extreme values to specify a trend of increasing size where none exists when the full range of variation is properly considered.

larged perspective—offering a different and often opposite interpretation—for this very situation of "trends" myopically depicted as "things" moving somewhere rather than as changes in variation of entire systems ("full houses").

Let us, then, following the procedure advocated throughout this book, portray the full range of variation through time in all three acts (Figure 23, based on data for first appearances of 377 species, supplied to me by Richard Norris of the Woods Hole Oceanographic Institution, and used by me in a technical paper published in 1988). Just as 0.400 hitting is not an independent "thing," but rather the right tail of the bell curve for batting averages, so too must we view the largest foram as an extreme value in a full distribution, not as an entity unto itself. When we consider the entire system, new modes of interpretation must be explored.

All traditional interpretations of Cope's Rule have been framed in terms of supposed evolutionary advantages for larger bodies. I mean, how else could one possibly proceed? Body size clearly increases as a general-

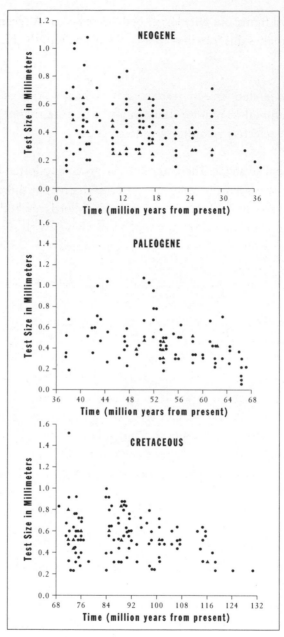

FIGURE 23

Size at first appearance for species of planktonic forams during their three evolutionary radiations. Note that species start small (right-hand side of each graph) in each interval—and that many species remain small while the range increases in each interval.

ity, so we must figure out why larger bodies are better. A recent article on Cope's Rule makes this "obvious" point clearly and bluntly (Hallam, 1990, page 264):

> Since phyletic size increase is such a widespread trend in the animal kingdom, there must be manifestly one or more selective advantages of larger size.

Traditional strategies then continue by proposing (often by speculation, or at least without consideration of alternatives) a short list of advantages that would lead natural selection to favor large bodies in most cases. Hallam continues from the quotation above (with reasons better suited to large multicellular animals than to forams):

> Among those proposed are an improved ability to capture prey or ward off predators; greater reproductive success; increased regulation of the internal environment; and increased heat regulation per unit volume.

Another recent article, titled "Body size, ecological dominance, and Cope's rule" (Brown and Maurer, 1986, page 250), proposes a most important benefit of all: "Presumably the ecological advantage of monopolizing resources provides the selective pressure that promotes evolution of greater size. Individuals of large size are favored by . . . natural selection, because they can dominate resources use and consequently leave more offspring than their smaller relatives."

I do confess to great discomfort when I see such words as "manifestly," or even "presumably," attached to conclusions stated without compelling logic or evidence (or subject to another interpretation simply not conceptualized by the author). I am reminded of the chilling line that Wilson attributed to Peirce (as quoted in chapter 2): "Let us not pretend to deny in our philosophy what we know in our hearts to be true." Such protestations of the "obvious" stymie thought; the non-obvious is so often true— and, when true, usually enormously interesting (if only for the power of breaking through old prejudices). Figure 22 is a myopically misleading picture with two "manifest" features that need not be true: an "obvious" evo-

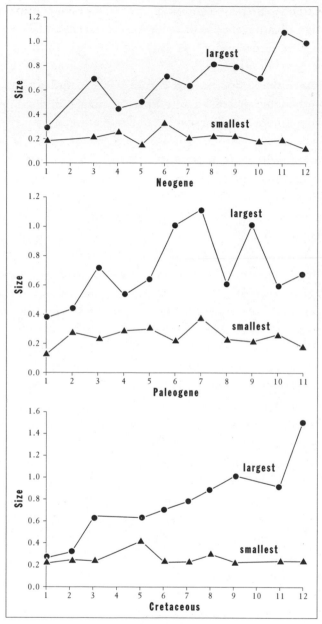

FIGURE 24

Although size of the largest species tends to increase in each of the three radiations of forams, the size of the smallest species either remains constant or decreases. Therefore the main trend is toward an increase in variation of size, not a directional movement.

lutionary trend to increased body size, with a "necessary" implication that the selective advantages of such an increase must cause the trend.

The full variation of Figure 23 shows an increase through time of the largest species, but no general trend for the entire lineage. Small species continue to exist and flourish (see Figure 24 for histories of smallest and largest on the same chart for each act). If we must talk of "trends" at all, should we not notice and emphasize the increasing *range* of variation in size during each act? At each of the three beginnings, evolution starts with a few founding lineages at small body size. The range then increases through time. Small species continue to flourish (and always constitute the largest fraction of species), while the range of size for all species expands. How can we say that larger size holds an absolute advantage, when most species remain small?

A supporter of traditional interpretations for Cope's Rule might reply, "Yes, I see your point about the continuity of small species. But some species get bigger, while none get smaller than the founding lineages. So some (at least statistical) advantage for large size must exist." Fine, except for one point and a key theme of this book: walls.

Remember the wall in the drunkard's walk. His random motion could accumulate in only one direction because he started at a wall that he couldn't penetrate. Remember the wall in 0.400 hitting. The best batter cannot penetrate the right wall of human limitation—so he must stay in the same spot, hands against the wall, while the mean player sneaks up upon him, leading to a reduction in his batting average for the same performance. Can we talk about a similar wall in the evolution of planktonic forams?

We now come to the curious point that makes this example so compelling—for we could not possibly imagine a more undeniable wall than the lower limit of body size for planktonic forams. I say this with some cynicism because this wall is entirely an artifact of an arbitrary human decision, not a dictate of nature at all. But what could be more clearly definable than an arbitrary decision of art?

Forams are nearly (or actually) microscopic. They cannot be collected by unaided visual inspection. Planktonic forams exist in multitudes within marine sediment. They are recovered for study by disaggregating the sediment and washing it through a stacked series of sieves, with decreasing mesh size from top to bottom. Thus the largest particles get trapped in the upper sieve, while particles smaller than the mesh of the lowest sieve get washed down the drain. Traditionally, although practices do vary somewhat among laboratories, the smallest sieve has a mesh size of 150 micrometers. If forams smaller than 150 micrometers exist (and they do), they end up in the sink and do not appear in our figures. A size of 150 micrometers therefore operates as a true left wall of minimal dimensions for the evolution of forams. If founding lineages begin near this left wall—and they do in all three acts—then nothing can get any smaller later in the act.

The existence of this left wall forces a reevaluation of the entire story. Need we postulate any more than the presence of this wall, and the beginning of each act in its vicinity, to render the apparent trend? Need we say anything about the putative advantages of large size at all? Only one direction of change lies open. Forams cannot get any smaller than their beginnings, but many species retain this initial size and continue to flourish. Others expand into the only open space.

We shouldn't deny an impetus to large size just because some species remain small. Perhaps just a few stragglers retain the starting dimensions, while most follow Cope's Rule for conventional reasons of "bigger is better." But two additional kinds of evidence strongly argue against this last possibility for viewing the play of planktonic forams as a story of repeated benefits for larger species.

First let us consider the history of size for the most appropriate measure of an "average" species in each act—for if this "average" tends to increase, then maybe we should view larger size as a property of the whole. In chapter 4, I listed three major statistics for calculating an average—mean, median, and mode—and I discussed occasions when one or the other cannot be deemed appropriate. In particular, means and medians can give false impressions in highly skewed distributions—for both these measures will be pulled strongly in the direction of skew (the mean more than the median), even if very few individuals occupy the extended tail of the skewed distribution. If this general statement seems overly abstract, recall the discussion of the bell curve for incomes—a strongly right-skewed distribution because a few Bill Gateses make a billion bucks a year, while a left wall stands at no income at all. Thus, a whole lot of folks have to be gathered between the left wall of zero and the mean family income of about $30,000 per year—while the right tail extends out almost forever to Gates and (a very limited) company.

The mean is a terrible measure for any vernacular notion of "average" or "central tendency" in such a highly skewed distribution, because the introduction of just one Bill Gates will pull the mean way to the right—for his billion dollars to the right of the mean counts as much as 100,000 people making $10,000 to the left of the mean. The mean of such a highly skewed distribution therefore moves far away from the peak of the most frequent value—and ends up on the bell curve's flank in the direction of skew (see Figure 7 for a graphic illustration and fuller discussion of this important principle). The median is not so badly pulled as the mean, but the median also ends up too far from the bell curve's peak and on the skewed flank of the distribution (again, see Figure 7 and the accompanying discussion).

This artifact severely distorts our interpretation of a figure like 22, where we want to read the steady rise of mean values as a sure sign of general increase in size (according to Cope's Rule) for the whole group of planktonic forams—whereas such a rise could also just indicate that a bell curve with an unchanging peak value has become progressively more right skewed with time. For highly skewed distributions, we therefore generally favor the third major measure of central tendency—the mode, or most common value (that is, the peak value of the bell curve itself).

I therefore divided the total range of variation attained within each act into ten equal intervals, and I plotted the interval occupied by the most species (I called this interval the "modal decade") for each of twelve equal time units in each of the three acts. I show my results in Figure 25. For this preferred measure of mode, we find no tendency whatsoever for increase in size through time in any act. (The Cretaceous act does increase a bit for the first three intervals, but then remains steady; the Paleogene act decreases in size of modal decade near the end; the Neogene act stays pretty steady throughout.) In other words, the most common of all sizes does not change substantially—either up or down—during any of the three episodes in the evolution of planktonic forams.

As a second, and clinching, piece of evidence, I dearly desired one more tabulation that was not available to me when I did my study in 1988—for

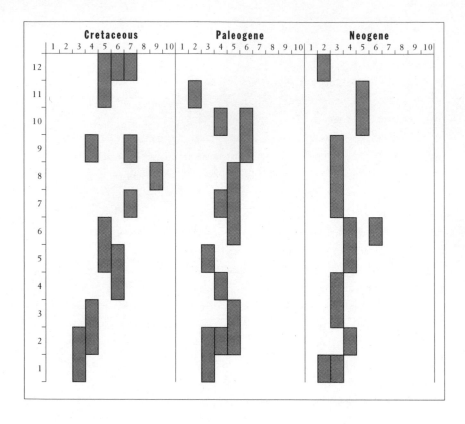

FIGURE 25

Change in the most frequently encountered size (small on left, large on right) during the three radiations of planktonic forams from oldest (bottom) to youngest (top). More than one shaded area at any time indicates equal frequencies for the shaded size classes. Note that, except for an initial size increase in the Cretaceous, the most common size shows no tendency to increase in any of the three radiations.

I did not know the actual sequence of ancestor-descendant pairings. We want to learn whether any tendency exists for descendant species to arise at a larger size than their immediate ancestors—for if decreases in size are as common as increases in known evolutionary transitions, then we surely cannot talk meaningfully about an "impetus" or "trend" to size increase, even if the size of the largest species, or the inappropriate mean value, gets larger as the full distribution becomes progressively right skewed through time.

My colleague Anthony J. Arnold, along with his associates D. C. Kelly

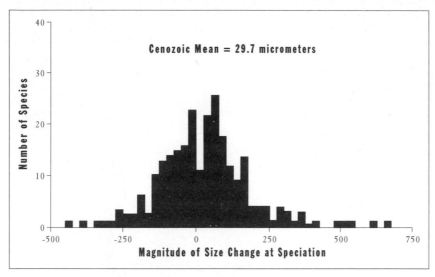

FIGURE 26

Evidence that no preference for increasing size exists in speciation events during the evolution of forams. Values greater than zero on the horizontal axis indicate size increase; values less than zero indicate size decrease in a speciation event. The normal distribution shows no preference for either increase or decrease during evolution.

and W. C. Parker of Florida State University, have now supplied the missing information. Using a remarkable set of data on known pairings of ancestors with descendants in 342 Cenozoic species of planktonic forams, they plotted the bell curve for differences in size between descendants and their immediate ancestors. A zero value in Figure 26 indicates that the descendant arose at the same size as the ancestor, a negative value marks a smaller descendant, and a positive value a larger descendant. The symmetrical, unskewed bell curve of Figure 26 proves that no tendency exists for either increase or decrease of size in the origin of new species in planktonic forams. A descendant is just as likely to arise at either a smaller or a larger size than its ancestor. Arnold, Kelly, and Parker (1995, page 206) state their clear conclusion: "There is no apparent . . . tendency to favor size increase; there is no strong indication of size-dependent longevity, and there is no indication of size dependence in speciation or extinction rates."

Yet the size of the largest species, and the mean value, do increase within each act. How shall we interpret this phenomenon if we must deny—as a proper examination of variation in the full system says we must—any overall tendency or general advantage to increasing size? Ironically, we seem to need an explanation precisely backwards from the usual claim (the "manifest" and "obvious" superiority of large size, now discredited). The entire phenomenon arises from three factors: (1) the existence of a left wall, a true lower limit to size in this case, set by the artifact of minimum mesh in conventional laboratory sieves; (2) the survival of small-bodied species alone (near the minimal size) after each episode of mass extinction, and the consequent beginning of each act only with species at the lower end of the size range; and (3) successful radiation and increase in number of species within each act, so that total diversity grows through time in each case.

Given these three conditions, we note an increase in size of the largest species only because founding species start at the left wall, and the range of size can therefore expand in only one direction. Size of the most common species (the modal decade) never changes, and descendants show no bias for arising at larger sizes than ancestors. But, during each act, the range of size expands in the only open direction by increase in the total number of species, a few of which (and only a few) become larger (while none can penetrate the left wall and get smaller). We can say only this for Cope's Rule: in cases with boundary conditions like the three listed above, extreme achievements in body size will move away from initial values near walls. Size increase, in other words, is really *random evolution away from small size, not directed evolution toward large size.*

Please understand that I am not depriving this story of great interest and importance, and I am not denying that the size of the largest species increases with time. I am saying that proper consideration of expanding variation within the full system, rather than myopic focus on mean or extreme values ("things moving somewhere"), forces us to reinterpret the case in a manner opposite to the usual reading. Under the conventional view we asked why selection favored large size. In the new interpretation we need to know (and we do not) why small-bodied species differentially survived in episodes of mass extinction to begin each new episode of evolution with just a few species at nearly minimal size. Everything else simply follows from this limited beginning and the group's expanding success.

The necessity (and fascination) of such an inverted interpretation in this case suggests that we might profitably reassess the entire phenomenon of Cope's Rule, one of the oldest "received truths" in paleontology and evolutionary theory. I do not doubt that some instances may be best explained under the old rubric of "things moving somewhere"—that is, as general increases in size for all or most lineages in a group as a result of selective advantage for larger bodies (and not as expansion of variation in a full system, misread as a trend in extreme values).

But a survey of all cases will surely alter our former certainty and will teach us to prefer full houses over abstracted averages or extremes if we wish to unravel both the phenomenology and the causality of evolutionary change in the fossil record. First of all, some venerable cases of Cope's Rule are pure artifacts of a myopic focus upon extreme values. For example, my colleague David Jablonski of the University of Chicago studied patterns of change in size for all genera of clams with fossil records spanning more than 4 million years in late Cretaceous sediments of the Gulf and Atlantic Coastal Plain of the United States. He found that thirty-three of fifty-eight genera followed what he called the "broad" (I would say inappropriate) sense of Cope's Rule, because the largest late representative exceeded the largest early representative in size. But he then found that, in twenty-two of these thirty-three genera, the size of the smallest species also decreased or remained stable in time. Thus, in at least two-thirds of the genera studied, "general" size increase records only our tendency to study the upper bound rather than the entire range. Jablonski concluded (1987, page 714) that "Cope's Rule is driven by an increase in variance rather than a simple directional trend in body sizes."

In other, more legitimate cases, increases in means or extremes occur, as in our story of planktonic forams, because lineages started near the left wall of a potential range in size and then filled available space as the number of species increased—in other words, a drift of means or extremes *away from small size,* rather than directed evolution of lineages *toward large size* (and remember that such a drift can occur within a regime of random change in size for each individual lineage—the "drunkard's walk" model).

In 1973, my colleague Steven Stanley of Johns Hopkins University published a marvelous, and now celebrated, paper to advance this important argument. He showed (see Figure 27, taken from his work) that groups beginning at small size, and constrained by a left wall near this starting point, will increase in mean or extreme size under a regime of random evolution within each species. He also advocated that we test his idea by looking for right-skewed distributions of size within entire systems, rather than by tracking mean or extreme values that falsely abstract such systems as single numbers. In a 1988 paper I suggested that we speak of "Stanley's Rule" when such an increase of means or extremes can best be explained by undirected evolution away from a starting point near a left wall. I would venture to guess (in fact I would wager substantial money on the proposition) that a large majority of lineages showing increase of body size for mean or extreme values (Cope's Rule in the broad sense) will properly be explained by Stanley's Rule of random evolution away from small size rather than by the conventional account of directed evolution toward selectively advantageous large size.

In this context, and to conclude the chapter, I was delighted to discover (when studying Cope's original formulation in order to write this part of my book) that Cope himself had grasped this better explanation "through a glass darkly." Cope did write a good deal about the phenomenon that would later be called "Cope's Rule" or even "Cope's Law." But he devoted much more attention to another putative law that he evidently regarded as much more important—his self-styled "law of the unspecialized" (Cope, 1896, pages 172–74).

This law states that founding members of highly successful lineages tend to be "unspecialized" in the sense that they can tolerate a wide range of habitats and climates, and that they do not possess complex and highly specific adaptations to narrow behaviors or modes of life (the peacock's tail or the koala's need to eat just one kind of eucalyptus leaf). With the proviso that such evolutionary laws have only majoritarian status (not exclusivity) in our complex and partly random world, Cope's law of the unspecialized has held up well, and would be endorsed by evolutionary biologists today.

Cope himself recognized—and not just as a passing observation, for he repeated the point several times—that such unspecialized lineages also tend to be small in body size (and even that small size favors unspecialized status). But he never made the full connection that becomes so obvi-

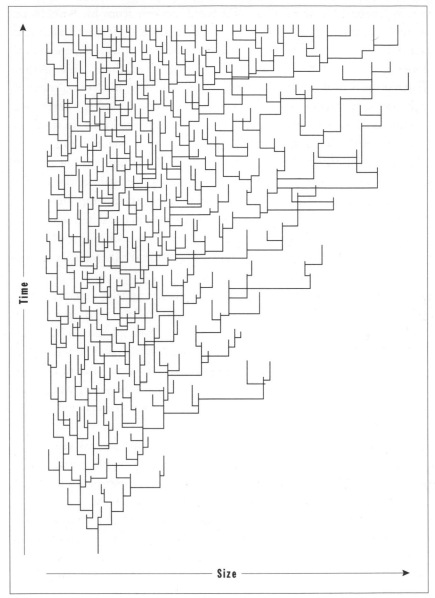

FIGURE 27
The increase in size of mean and extreme values within a branching evolutionary series arises in this case only as a function of origin near a lower limit, or left wall, in size.

ous in the light of this book's central theme: maybe Cope's now more famous law of increasing body size only arises as a noncausal side consequence of Cope's other law of the unspecialized. Major lineages tend to
be founded by species of unspecialized anatomy and behavior. Unspecialized species also tend to be small in body size. Cope's Rule of increasing size is therefore an artifact of these small beginnings near a left wall
for the founders of major lineages. Cope never made all the connections,
but we should record and honor his words:

> The "Doctrine of the Unspecialized" . . . describes the
> fact that the highly developed, or specialized types of
> one geological period have not been the parents of the
> types of succeeding periods, but that the descent has been
> derived from the less specialized of preceding ages. . . .
> The validity of this law is due to the fact that the spe-
> cialized types of all periods have been generally inca-
> pable of adaptation to the changed conditions which
> characterized the advent of new periods. . . . Such
> changes have been often especially severe in their effects
> on species of large size, which required food in great
> quantities. . . . Animals of omnivorous food-habits would
> survive where those which required special foods would
> die. Species of small size would survive a scarcity of food,
> while large ones would perish. It is true . . . that the lines
> of descent of Mammalia have originated or been contin-
> ued through forms of small size. The same is true of all
> other Vertebrata.

·14·

The Power of the Modal Bacter,
or Why the Tail Can't Wag
the Dog

An Epitome of the Argument

I believe that the most knowledgeable students of life's history have always sensed the failure of the fossil record to supply the most desired ingredient of Western comfort: a clear signal of progress measured as some form of steadily increasing complexity for life as a whole through time. The basic evidence cannot support such a view, for simple forms still predominate in most environments, as they always have. Faced with this undeniable fact, supporters of progress (that is, nearly all of us throughout the history of evolutionary thought) have shifted criteria and ended up grasping at straws. (The altered criterion may not have struck the graspers as such a thin reed, for one must first internalize the argument of this book—trends as changes in variation rather than things moving some-where—to recognize the weakness.) In short, graspers for progress have looked exclusively at the history of the most complex organism through time—a myopic focus on extreme values only—and have used the in-creasing complexity of the most complex as a false surrogate for progress of the whole (again, see this book's opening example and Figure 1 for a striking case). But this argument is illogical and has always disturbed the most critical consumers.

Thus, James Dwight Dana, America's greatest naturalist in Darwin's era (at least after Agassiz's death), and a soul mate to Darwin in their re-markably parallel careers (both went on long sea voyages in their youth, and both became fascinated with coral reefs and the taxonomy of crus-taceans), used this criterion when he finally converted to evolution in the mid-1870s. Dana's primary commitment to progress as the definition of life's organization held firm throughout his career, and in his personal transition from creationism to evolution. But Dana could validate progress only by looking at the history of extremes—"the grand fact that the sys-tem of life began in the simple sea-plant and the lower forms of animals, and ended in man" (Dana, 1876, page 593). Julian Huxley, grandson of Thomas Henry, sensed the same unease, but could think of no other cri-terion in 1959 (as quoted at the beginning of chapter 12). When Darwin's grandson challenged him to defend progress in the light of so many well-adapted but anatomically simplified parasites, Julian Huxley replied, "I mean a higher degree of organization in general, as shown by the upper level attained." But the "upper level attained" (the extreme organism at the right tail) is not a measure of "organization in general"—and Hux-ley's defense is illogical.

In debunking this conventional argument for progress in the history of life, I reach the crux of this book (although I will not disparage anyone who regards baseball as equal in importance to life's history, and there-fore views the correct interpretation of 0.400 hitting as more vital to Amer-ican life than understanding the central themes of 3.5 billion years in biological time)! Yet I can summarize my argument against progress in the history of life in just seven statements condensed into a few pages. I do not mean to be capricious or disrespectful in this brevity. If I have done my job in the rest of this book, I have already set the background and argument with sufficient thoroughness—so this focal application at grandest scale should follow quickly with just a few reminders and way stations for the new context.

I do not challenge the statement that the most complex creature has tended to increase in elaboration through time, but I fervently deny that this limited little fact can provide an argument for general progress as a defining thrust of life's history. Such a grandiose claim represents a ludicrous case of the tail wagging the dog, or the invalid elevation of a small and epiphenomenal consequence into a major and controlling cause.

I shall present, in seven arguments, my best sense of a proper case based on the history of expanding variation away from a beginning left wall. I shall then provide extended commentary for three of the statements that are particularly vital, and most generally misunderstood or unappreciated. Please note that the entire sequence of statements for life as a whole follows exactly the same logic, and postulates the same causes, as my previous story (at smallest scale) about the evolution of planktonic forams.

1. *Life's necessary beginning at the left wall.*   The earth is about 4.5 billion years old. Life, as recorded in the fossil record, originated at least 3.5 billion years ago, and probably not much earlier because the earth passed through a molten period that ended about 3.8 billion years ago (the age of the oldest rocks). Life presumably began in primeval oceans as a result of sequential chemical reactions based on original constituents of atmospheres and oceans, and regulated by principles of physics for self-organizing systems. (The "primeval soup" has long been a catchword for oceans teeming with appropriate organic compounds prior to the origin of life.) In any case, we may specify as a "left wall" the minimal complexity of life under these conditions of spontaneous origin. (As a paleontologist, I like to think of this wall as the lower limit of "conceivable, preservable complexity" in the fossil record.) For reasons of physics and chemistry, life had to begin right next to the left wall of minimal complexity—as a microscopic blob. You cannot begin by precipitating a lion out of the primeval soup.

2. *Stability throughout time of the initial bacterial mode.*   If we are particularly parochial in our concern for multicellular creatures, we place the major division in life between plants and animals (as the Book of Genesis

does in both creation myths of chapters 1 and 2). If we are more ecumenical, we generally place the division between unicellular and multicellular forms. But most professional biologists would argue that the break of maximal profundity occurs within the unicells, separating the prokaryotes (or cells without organelles—no nuclei, no chromosomes, no mitochondria, no chloroplasts) from the eukaryotes (organisms like amoebae and paramecia, with all the complex parts contained in the cells of multicellular organisms). Prokaryotes include the amazingly diverse groups collectively known as "bacteria," and also the so-called "blue-green algae," which are little more than photosynthesizing bacteria, and are now generally known as Cyanobacteria.

All the earliest forms of life in the fossil record are prokaryotes—or, loosely, "bacteria." In fact, more than half the history of life is a tale of bacteria only. In terms of preservable anatomy in the fossil record, bacteria lie right next to the left wall of minimal conceivable complexity. Life therefore began with a bacterial mode (see Figure 28). Life still maintains a bacterial mode in the same position. So it was in the beginning, is now, and ever shall be—at least until the sun explodes and dooms the planet. How, then, using the proper criterion of variation in life's full house, can we possibly argue that progress provides a central defining thrust to evolution if complexity's mode has never changed? (Life's *mean* complexity may have increased, but see chapter 4 for a discussion of why means are inappropriate, and modes proper, as measures of central tendency in

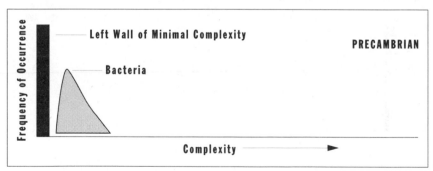

FIGURE 28
Life begins necessarily near the left wall of minimal complexity, and the bacterial mode soon develops.

strongly skewed distributions.) The modal bacter of this chapter's title has been life's constant paradigm of success.

3. *Life's successful expansion must form an increasingly right-skewed distribution.* Life had to begin next to the left wall of minimal complexity (see statement 1). As life diversified, only one direction stood open for expansion. Nothing much could move left and fit between the initial bacterial mode and the left wall. The bacterial mode itself has maintained its initial position and grown continually in height (see Figure 29). Since space remains available away from the left wall and toward the direction of greater complexity, new species occasionally wander into this previously unoccupied domain, giving the bell curve of complexity for all species a right skew, with capacity for increased skewing through time.

4. *The myopia of characterizing a full distribution by an extreme item at one tail.* Considering life's full house of Figure 29, the only conceivable ar-

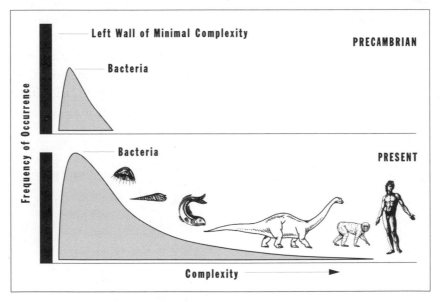

FIGURE 29
The frequency distribution for life's complexity becomes increasingly right skewed through time, but the bacterial mode never alters.

gument for general progress must postulate that an expanding right tail demonstrates a predictable upward thrust of the whole. But such a claim only embodies the silly spectacle of a small tail wagging a large dog. (We have generally failed to grasp the evident absurdity because we have not visualized the dog properly; rather, in a move that recalls the Cheshire Cat of Wonderland, identified only by its smile, we have characterized the entire dog by its tail alone.)

A claim for general progress based on the right tail alone is absurd for two primary reasons: First, the tail is small and occupied by only a tiny percentage of species (more than 80 percent of multicellular animal species are arthropods, and we generally regard almost all members of this phylum as primitive and nonprogressive). Second, the occupants of the extreme right edge through time do not form an evolutionary sequence, but rather a motley series of disparate forms that have tumbled into this position, one after the other. Such a sequence through time might read: bacterium, eukaryotic cell, marine alga, jellyfish, trilobite, nautiloid, placoderm fish, dinosaur, saber-toothed cat, and *Homo sapiens*. Beyond the first two transitions, not a single form in this sequence can possibly be a direct ancestor of the next in line.

5. *Causality resides at the wall and in the spread of variation; the right tail is a consequence, not a cause.* The development of life's bell curve for complexity through time (Figures 28 and 29) does not represent a fully random phenomenon (though random elements play an important role). Two important causal influences shape the curve and its changes—but neither influence includes any statement about conventional progress. The two major causes are, first, necessary origin at the left wall of minimal complexity; and, second, increase of numbers and kinds, with predictable development of a right-skewed distribution. Given this point of origin at a wall and subsequent increase in variation, the right tail almost had to develop and extend. But this expansion of the right tail—the only (and myopic) source for any potential claim about progress—is an epiphenomenon and a side consequence of the two causes listed above, not a fundamental thrust produced by the superiority of complex forms under natural selection. In fact, as the paradigm of the drunkard's walk illustrates, such an extension of the right tail will occur in a regime of entirely random mo-

tion for each item, so long as the system begins at a wall. Thus, as the drunkard's walk shows in theory, and the evolution of planktonic forams confirms in fact, the expanding right tail of life's complexity may arise from random motion among all lineages. The vaunted progress of life is really *random motion away from simple beginnings,* not *directed impetus toward inherently advantageous complexity.*

6. *The only promising way to smuggle progress back into such a system is logically possible, but empirically false at high probability.*   My argument for the whole system is sound: from a necessary beginning at the left wall, random motion of all items in a growing system will produce an increasingly right-skewed distribution. Thus, and with powerful irony, the most venerable evidence for general progress—the increasing complexity of the most complex—becomes a passive consequence of growth in a system with no directional bias whatever in the motion of its components.

But one potential (though much vitiated) argument for general progress remains. The entire system is free to vary only in the direction of greater complexity from an initial position next to the left wall. But what about a smaller lineage that begins at some intermediary position with freedom to expand in *either* direction (the first living thing starts at the left wall, but the first mammal, or the first seed plant, or the first clam, begins in the middle and its descendants can move toward either tail). If we studied all the smaller lineages free to vary in any direction, perhaps we would then detect a clear bias for net movement to the right, or toward greater complexity. If we found such a bias, we could legitimately speak of a general trend to greater complexity in the evolutionary history of lineages. (This more subtle position would still not explain the general pattern of Figure 29, which would still arise as a consequence of random motion in a growing system constrained to begin at the left wall. But a rightward bias in individual lineages would function as a "booster" or "helpmate" in the general production of right skew. The entire system would then be built by two components: random motion from the left wall, and a rightward bias in individual lineages—and the second component would provide an argument for general progress.)

The logic of this argument is sound, but two strong reasons suggest (though not all the evidence is yet in) that the proposition is empirically false. (I shall summarize the two reasons here and provide more details later in this chapter.) First, while I know of no proven bias for rightward motion under natural selection—a mechanism that yields only local adaptation to changing environments, not general progress—a good case can be made for leftward bias because parasitism is such a common evolutionary strategy, and parasites tend to be anatomically more simplified than their free-living ancestors. (Ironically, then, the full system of increasing right skew for the whole might actually be built with a slight bias toward *decreased* complexity in individual lineages!) Second, several paleontologists are now studying this issue directly by trying to quantify the elusive notion of progress and then tracing the changing spread of their measure in the history of individual lineages. Only a few studies have been completed so far, but current results show no rightward bias, and therefore no tendency to progress in individual lineages.

7. *Even a parochial decision to focus on the right tail alone will not yield the one, most truly desired conclusion, the psychological impetus to our yearning for general progress—that is, the predictable and sensible evolution to domination of a creature like us, endowed with consciousness.* We might adopt a position of substantial retreat from an original claim for general progress, but still a bastion of defense for what really matters to us. That is, we might say, "Okay, you win. I understand your point that the evidence of supposed progress, the increasing right skew of life's bell curve, is only an epiphenomenal tail that cannot wag the entire dog—and that life's full house has never moved from its modal position. But I am allowed to be parochial. The right tail may be small and epiphenomenal, but I love the right tail because I dwell at its end—and I want to focus on the right tail alone because this little epiphenomenon is all that matters to me. Even you admit that the right tail had to arise, so long as life expanded. So the right tail had to develop and grow—and had to produce, at its apogee, something like me. I therefore remain the modern equivalent of the apple of God's eye: the predictably most complex creature that ever lived."

Wrong again, even for this pitifully restricted claim (after advancing an initial argument for intrinsic directionality in the basic causal thrust of all evolution). The right tail had to exist, but the actual composition of creatures on the tail is utterly unpredictable, partly random, and entirely contingent—not at all foreordained by the mechanisms of evolution. If we could replay the game of life again and again, always starting at the left wall and expanding thereafter in diversity, we would get a right tail almost every time, but the inhabitants of this region of greatest complexity would be wildly and unpredictably different in each rendition—and the vast majority of replays would never produce (on the finite scale of a planet's lifetime) a creature with self-consciousness. Humans are here by the luck of the draw, not the inevitability of life's direction or evolution's mechanism.

In any case, little tails, no tails, or whoever occupies the tails, the outstanding feature of life's history has been the stability of its bacterial mode over billions of years!

The Multifariousness of the Modal Bacter

My interest in paleontology began in a childhood fascination with dinosaurs. I spent a substantial part of my youth reading the modest literature then available for children on the history of life. I well remember the invariant scheme used to divide the fossil record into a series of "ages" representing the progress that supposedly marked the march of evolution: the "Age of Invertebrates," followed by the Age of Fishes, Reptiles, Mammals, and, finally, with all the parochiality of the engendered language then current, the "Age of Man."

I have watched various reforms in this system during the past forty years (though see chapter 2, for persisting use of the old scheme). The language police, of course, would never allow an Age of Man anymore, so we could, at best and with more inclusive generosity, now specify an "age of humans" or an "age of self-consciousness." But we have also come to recognize, with even further inclusive generosity, that one species of mammals, despite our unbounded success, cannot speak adequately for the whole. Some enlightened folks have even recognized that an "age of mammals" doesn't specify sufficient equity—especially since mammals form a small group of some four thousand species, while nearly a million species of multicellular animals have been formally named. Since more than 80 percent of these million are arthropods, and since the great majority of arthropods are insects, these same enlightened people tend to label modern times as the "age of arthropods."

Fair enough, if we wish to honor multicellular creatures—but we are still not free of the parochialism of our scale. If we must characterize a whole by a representative part, we certainly should honor life's constant mode. We live now in the "Age of Bacteria." Our planet has always been in the "Age of Bacteria," ever since the first fossils—bacteria, of course—were entombed in rocks more than three and a half billion years ago.

On any possible, reasonable, or fair criterion, bacteria are—and always have been—the dominant forms of life on earth. Our failure to grasp this most evident of biological facts arises in part from the blindness of our arrogance, but also, in large measure, as an effect of scale. We are so accustomed to viewing phenomena of our scale—sizes measured in feet and ages in decades—as typical of nature. Individual bacteria lie beneath our vision and may live no longer than the time I take to eat lunch, or my grandfather spent with his evening cigar. But then, who knows? To a bacterium, human bodies might appear as widely dispersed, effectively eternal (or at least geological), massive mountains, fit for all forms of exploitation, and fraught with little danger unless a bolus of imported penicillin strikes at some of the nasty brethren.

Consider just some of the criteria for bacterial domination:

TIME. I have already mentioned the persistence of bacterial rule. The fossil record of life begins with bacteria, some 3.5 to 3.6 billion years ago. About half the history of life later, the more elaborate eukaryotic cell makes a first appearance in the fossil record—about 1.8 to 1.9 billion years ago by best current evidence. The first multicellular creatures—marine algae—enter the stage soon afterward, but these organisms bear no genealogical relationship to our primary (if admittedly parochial) interest in this book: the history of animal life. The first multicellular animals do not enter the fossil record until about 580 million years ago—that is, after about five-sixths of life's history had already passed. Bacteria have been the stayers and keepers of life's history.

Moreover, bacteria do not record their history of Precambrian domination as invisible dots in rocks. Rather, they shaped their environments, and left their sedimentary records, in highly visible form—even though

FIGURE 30
Modern stromatolites—layers of sediment trapped and bound by prokaryotic cells.

no multicellular animals then lived to view the effect. The fossil record of ancient bacteria consists largely of stromatolites—complexly concentric and laminated layers, often looking like a head of cabbage in cross-section (see Figure 30). These sizable structures are not bacteria themselves, but layers of sediments trapped and bound by mats of bacterial cells. Most stromatolites formed near the tide lines, and were constantly desiccated and regrown during fluctuations in sea level—thus leading to large, vertical piles of wavy layers. Stromatolites still exist, but now can form only in un-usual environments devoid of the multicellular animals that happily feed on such organisms and therefore prevent their formation in most places. But no potential feeders lived in these early years, during most of life's his-tory, and stromatolites must have covered appropriate habitats through-out the planet.

INDESTRUCTIBILITY. Let us make a quick bow to the flip side of such long domination—to the future prospects that match such a distinguished and persistent past. Bacteria have occupied life's mode from the very be-ginning, and I cannot imagine a change of status, even under any con-ceivable new regime that human ingenuity might someday impose upon our planet. Bacteria exist in such overwhelming number, and such un-paralleled variety; they live in such a wide range of environments, and work in so many unmatched modes of metabolism. Our shenanigans, nu-clear and otherwise, might easily lead to our own destruction in the fore-seeable future. We might take most of the large terrestrial vertebrates with us—a few thousand species at most. We surely cannot extirpate 500,000 species of beetles, though we might make a significant dent. I doubt that we could ever substantially touch bacterial diversity. The modal organisms cannot be nuked into oblivion, or very much affected by any of our considerable conceivable malfeasances.

TAXONOMY. The history of classification for the basic groups of life is one long tale of decreasing parochialism and growing recognition of the diversity and importance of single-celled organisms, and other "lower" creatures. Most of Western history favored the biblically sanctioned twofold division of organisms into plants and animals (with a third realm for all inorganic substances—leading to the old taxonomy of "animal, vegetable, or mineral" in such venerable games as Twenty Questions). This twofold division produced a host of practical consequences, including the separation of biological research into two academic departments and tra-ditions of study: zoology and botany. Under this system, all single-celled organisms had to fall into one camp or the other, however uncomfortably, and however tight the shove of the shoehorn. Thus, paramecia and amoe-bae became animals because they move and ingest food. Photosynthesiz-ing unicells, of course, became plants. But what about photosynthesizers with mobility? And, above all, what about the prokaryotic bacteria, which bear no key feature suggesting either allocation? But since bacteria have a strong cell wall, and because many species are photosynthetic, bacteria fell into the domain of botany. To this day, we still talk about the bacterial "flora" of our guts.

By the time I entered high school in the mid-1950s, expansion and enlightenment had proceeded far enough to acknowledge that unicells could not be so divided by criteria of the multicellular world, and that single-celled organisms probably deserved a separate kingdom of their own, usually called Protista.

Twelve years later, as I left graduate school, even greater respect for the unicells had led to further proliferation at the "lower" end. A "five kingdom" system was now all the rage (and has since become canonical in textbooks)—with the three multicellular kingdoms of plants, fungi, and animals in a top layer (representing, loosely, production, decomposition, and ingestion as basic modes of life); the eukaryotic unicells, or Kingdom Protista, in a middle layer; and the prokaryotic unicells, or Kingdom Monera, representing bacteria and "blue-green algae," on a bottom rung. Most proponents of this system recognized the gap between prokaryotic and eukaryotic organization—that is, the transition from Monera to Protista—as the fundamental division within life, thus finally granting bacteria their measure of independent respect, if only as a bottom tier.

A decade later, starting in the mid-1970s, development of techniques for sequencing the genetic code finally gave us a key for mapping evolutionary relationships among bacterial lineages. (We know how to use anatomy for drawing genealogical trees of multicellular creatures more familiar to us—so we employ the internal skeleton of vertebrates, the external carapace of arthropods, and the multiplated test and radial symmetry of echinoderms to identify major evolutionary groups. But we are so ignorant of the bacterial world that we couldn't identify proper genealogical divisions—and we therefore tended to dump all bacteria together into a bag of little unicellular blobs, rods, and spirals. Yet we should have suspected deep divisions, far more extensive than those separating lines of multicellular animals—if only because bacteria have inhabited the planet for so long.)

As nucleotide sequences began to accumulate for key segments of bacterial genomes, a fascinating and unsuspected pattern emerged—and has grown ever stronger with passing years and further accumulation of evidence. This group of supposed primitives, once shoved into one small bag for their limited range of overt anatomical diversity, actually includes two great divisions, each far larger in scope (in terms of genomic distinction and variety) than all three multicellular kingdoms (plants, animals, and fungi) combined! Moreover, one of these divisions seemed to gather together, into one grand sibship, most of the bacteria living in odd environments and working by peculiar metabolisms under extreme conditions (often in the absence of oxygen) that may have flourished early in the earth's history—the methanogens, or methane producers; the tolerators of high salinities, the halophiles; and the thrivers at temperatures around the boiling point of water, the thermophiles.

These first accurate genealogical maps led to the apparently inescapable conclusion that two grand kingdoms, or domains, must be recognized within the old kingdom Monera—Bacteria for most conventional forms that come to mind when we contemplate this category (the photosynthesizing blue-greens, the gut bacteria, the organisms that cause human diseases and therefore become "germs" in our vernacular); and Archaea for the newly recognized coherence of oddballs. By contrast, all eukaryotic organisms, the three multicellular kingdoms as well as all unicellular eukaryotes, belong to a third great evolutionary domain, the Eucarya.

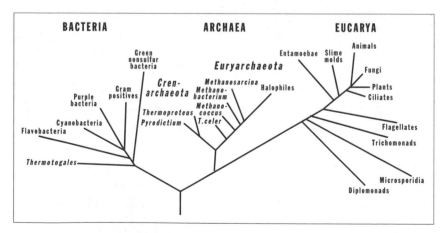

FIGURE 31

Life's evolutionary tree, showing two prokaryotic domains and only one eukaryotic domain, with plants, animals, and fungi as small twigs at an extreme of the eukaryotic domain.

The accompanying chart (Figure 31), from the work of Carl Woese, our greatest pioneer in this new constitution of life, says it all, with the maximally stunning device of a revolutionary picture. We now have a system of three grand evolutionary domains—Bacteria, Archaea, and Eucarya—and two of the three consist entirely of prokaryotes: that is, "bacteria" in the vernacular, the inhabitants of life's constant mode. Once we place two-thirds of evolutionary diversity at life's mode, we have much less trouble grasping the centrality of this location, and the constant domination of life by bacteria. For example, the domain of Bacteria, as presently defined, contains eleven major subdivisions, and the genetic distance between any pair is at least equal to the average separation between eukaryotic kingdoms such as plants and animals (Fuhrman, McCallum, and Davis, 1992).

Note, by contrast and in closing, the restricted domain of all three multicellular kingdoms. On this genealogical chart for all life, the three multicellular kingdoms form three little twigs on the bush of just one among three grand domains of life. Quite a change in one generation—from my parents' learning that everything living must be animal or vegetable, to the icon of my mature years: the kingdoms Animalia and Plantae as two little twigs amid a plethora of other branches on one of three bushes—with both other bushes growing bacteria, and only bacteria, all over.

UBIQUITY. The taxonomic criterion (Figure 31), while impressive, does not guarantee bacterial domination—and for a definite reason common to all genealogical schemes. Bacteria form the root of life's entire tree. For the first 2 billion years or so, about half of life's full history, bacteria alone built the tree of life. Therefore all multicellular creatures, as late arrivers, can only inhabit some topmost branches; the roots and trunk must be exclusively bacterial. This geometry does not make the case for calling our modern world an "Age of Bacteria" because the roots and trunk might now be atrophied, with only the multicellular branches flourishing. We need to show not only that bacteria build most of life's tree, but also that these bacterial foundations remain strong, healthy, vigorous, and fully supportive of the minor superstructure called multicellular life. Bacteria, indeed, have retained their predominant position, and hold sway not only by virtue of a long and illustrious history, but also for abundant reasons of contemporary vigor. Consider two aspects of ubiquity:

1. *Numbers.* Bacteria inhabit effectively every place suitable for the existence of life. Mother told you, after all, that bacterial "germs" require constant vigilance to combat their ubiquity in every breath and every mouthful—and the vast majority of bacteria are benign or irrelevant to us, not harmful agents of disease. One fact will suffice: during the course of life, the number of *E. coli* in the gut of each human being far exceeds the total number of people that now live and have ever inhabited the earth. (And *E. coli* is only one species in the normal gut "flora" of all humans.)

Numerical estimates, admittedly imprecise, are a stock in trade of all popular writing on bacteria. The *Encyclopaedia Britannica* tells us that bacteria live by "billions in a gram of rich garden soil and millions in one drop of saliva." Sagan and Margulis (1988, page 4) write that "human skin harbors some 100,000 microbes per square centimeter" ("microbes" includes nonbacterial unicells, but the overwhelming majority of "microbes" are bacteria); and that "one spoonful of high quality soil contains about 10 trillion bacteria." I was particularly impressed with this statement about our colonial status (Margulis and Sagan, 1986): "Fully ten percent of our own dry body weight consists of bacteria, some of which, although they are not a congenital part of our bodies, we can't live without."

2. *Places.* Since the temperature tolerance and metabolic ranges of bacteria so far exceed the scope of all other organisms, bacteria live in all habitats accessible to any form of life, while the edges of life's toleration are almost exclusively bacterial—from the coldest puddles on glaciers, to the hot springs of Yellowstone Park, to oceanic vents where water issues from the earth's interior at 480°F (still below the boiling point at the high pressures of oceanic bottoms). At temperatures greater than 160°F, all life is bacterial. I shall say more in the following pages about new information on bacteria of the open oceans and the earth's interior, but even conventional data from terrestrial environments prove the point. *Thermophila acidophilum* thrives at 140°F, and at a pH of 1 or 2, the acidity of concentrated sulfuric acid. This species, found on the surface of burning coals, and in the hot springs of Yellowstone Park, freezes to death below 100°F.

UTILITY. Importance for human life forms the most parochial of criteria for assessing the role of any organism in the history and constitution of life—though the conventional case for bacteria proceeds largely in this mode. I will therefore expand a bit toward utility (or at least "intrinsicness") for all of life, and even for the earth.

1. *Historical.* Oxygen, the most essential constituent of the atmosphere for human needs, now maintains itself primarily through release by multicellular plants in the process of photosynthesis. The earth's original atmosphere apparently contained little or no free oxygen, and this otherwise unlikely element both arose historically, and is now maintained, by the action of organisms. Plants may provide the major input today, but oxygen started to accumulate in the atmosphere about 2 billion years ago, substantially before the evolution of multicellular plant life. Bacterial photosynthesis supplied the atmosphere's original oxygen (and, in concert with multicellular plants, continues to act as a major source of resupply today).

But even if plants release most of today's oxygen, the source of resupply remains, ultimately and evolutionarily, bacterial. The photosynthetic organelle of the eukaryotic cell—the chloroplast—is, by ancestry, a photosynthesizing bacterium. According to an elegant and persuasive notion—the endosymbiotic theory for the origin of the eukaryotic cell—several organelles of eukaryotes arose by greater coordination and integration of an original symbiotic assemblage of prokaryotic cells. In this sense, the eukaryotic cell began as a colony, and each unit of our own body can be traced to such a cooperative beginning.

The case has been made persuasively only for the mitochondrion—the "energy factory" of all cells—and the chloroplast—the photosynthetic organelle—though some proponents extend the argument more generally to cilia (seen as descendants of spirochete bacteria) and other parts of cells. The evidence seems entirely convincing for mitochondria and chloroplasts: both are about the same size as bacteria (prokaryotes are substantially smaller than eukaryotes, so several bacterial cells easily fit inside a eukaryote); they look and function like bacteria; they have their own DNA programs (small because most genetic material has, through evolutionary time, been transferred to the nucleus)—all indicating ancestral status as independent organisms. Thus, even today, atmospheric oxygen is a bacterial product—released either directly by bacterial photosynthesis, or by bacterial descendants in eukaryotic cells.

Bacterial symbiosis—with bacteria remaining as coherent creatures, taxonomically independent if ecologically dependent, and not fully incorporated like mitochondria and chloroplasts—is a vital and potent phenomenon in many of life's central processes and balances. We could not digest and absorb food properly without our gut "flora." Grazing animals, cattle and their relatives, depend upon bacteria in their complex, quadripartite stomachs to digest grasses in the process of rumination. About 30 percent of atmospheric methane can be traced to the action of methanogenic bacteria in the guts of ruminants, largely released into the atmosphere—how else to say it—by belches and farts. (The most cultured and distinguished British ecologist, G. Evelyn Hutchinson, once published a famous calculation on the substantial contributions to atmospheric methane made by the flatulence of domestic cattle. Sagan and Margulis [1988, page 113] advance the "semiserious suggestion that the primary function served by large mammals is the equitable distribution of methane gas throughout the biosphere.")

In another symbiosis essential to human agriculture, plants need nitrogen as an essential soil nutrient, but cannot use the ubiquitous free nitrogen of our atmosphere. This nitrogen is "fixed," or chemically converted into usable form, by the action of bacteria like *Rhizobium,* living symbiotically in bulbous growths on the roots of leguminous plants.

Some symbioses are eerie in their complexity and almost gory precision. Nealson (1991) documents the story of a nematode (a tiny roundworm) parasitic upon insects and potentially useful as a biological control upon pests. The nematode enters the insect's mouth, anus, or spiracle (breathing organ) and migrates into the hemocoel (or blood cavity). There the nematode ejects millions of bacterial symbionts from its own intestine into the insect's circulatory system. These bacteria, though harmless to the nematode, kill the insect within hours. (Bacteria need the nematode to feast upon the insect because bacteria entering by themselves never reach the hemocoel and therefore do not attack the insect.) The dead insect becomes bioluminescent (another consequence of bacterial action) and darkly pigmented, but does not putrefy (perhaps because the nematode also releases antibiotics that kill other bacteria but leave their own symbionts harmless).

The pigment and glow then attract other nematodes to the insectan feast. The nematodes grow and reproduce by eating the insect; they also take on the helpful bacteria as symbionts. This source can yield up to 500,000 nematodes per gram of infected insects.

The recent discovery of the remarkable deep-sea "vent faunas," at zones of effusion for hot, mineral-laden waters from the earth's interior to the ocean floor, has provided another striking case of bacterial necessity and symbiosis. An old saw of biological pedagogy (I well remember the phrase emblazoned on the chapter heading of my junior high school textbook) proclaims, "All energy for biological processes comes ultimately from the sun." (I remember the pains that teachers took to trace even the most indirect pathways to a solar source—worms on the sea bottom eating decomposed bodies of fishes, which had fed on other fishes in shallow waters, with the little fishes eating shrimp, shrimp eating copepods, copepods ingesting algal cells, and algal cells growing by photosynthesis from that ultimate solar source.)

The vent faunas provide the first exception to this venerable rule, for their ultimate source of energy comes from the heat of the earth's interior (which warms the emerging waters, contributes to the solubility of minerals, and so on). Bacteria form the base of this unique and independent food chain—mostly sulfur-oxidizing forms that can convert the minerals of emerging waters into metabolically useful form. Some rift organisms form amazing symbiotic associations with these bacteria. The largest animal of this fauna, the vestimentiferan worm *Riftia pachyptila,* grows to several feet in length, but has no mouth, gut, or anus. This creature is so morphologically simplified that taxonomists have still not been able to determine its zoological affinity with confidence (current opinion favors a status within a small group of marine worms, the phylum Pogonophora). *Riftia* does contain a large and highly vascularized organ called the trophosome, filled with specialized cells (bacteriocytes) that house the symbiotic sulfur bacteria. Up to 35 percent of the trophosome's weight consists of these bacteria (Vetter, 1991).

2. *Current.* As discussed above, bacteria produced our atmospheric oxygen, fix nitrogen in our soil, facilitate the rumination of grazing animals, and build the food web of the only nonsolar ecosystem on our planet. We

could also compile a long list of more parochial uses for particularly human needs and pleasures: the degradation of sewage to nutrients suitable for plant growth; the possible dispersion of oceanic oil spills; the production of cheeses, buttermilk, and yogurt by fermentation (we make most alcoholic drinks by fermentation of eukaryotic yeasts); the bacterial production of vinegar from alcohol, and of MSG from sugars.

More generally, bacteria (along with fungi) are the main reducers of dead organic matter, and thus act as one of the two major links in the fundamental ecological cycle of production (plant photosynthesis and, come to think of it, bacterial photosynthesis as well) and reduction to useful form for renewed production. (The ingesting animals are just a little blip upon this basic cycle; the biosphere could do very well without them.) Sagan and Margulis write in conclusion (1988, pages 4–5):

> All of the elements crucial to global life—oxygen, nitrogen, phosphorus, sulfur, carbon—return to a usable form through the intervention of microbes. . . . Ecology is based on the restorative decomposition of microbes and molds, acting on plants and animals after they have died to return their valuable chemical nutrients to the total living system of life on earth.

NEW DATA ON BACTERIAL BIOMASS. This range of bacterial habitation and necessary activity certainly makes a good case for domination of life by the modal bacter. But one claim, formerly regarded as wildly improbable but now quite plausible, if still unproven, would really clinch the argument. We may grant bacteria all the above, but surely the main weight of life rests upon eukaryotes, particularly upon the wood of our forests. Another truism in biology has long proclaimed that the highest percentage of the earth's biomass—pure weight of organically produced matter—must lie in the wood of plants. Bacteria may be ubiquitous and present in nearly uncountable numbers, but they are awfully light, and you need several gazillion to equal the weight of even a small tree. So how could bacterial biomass even come close to that of the displacing and superseding eukaryotes? But new discoveries in the open oceans and the earth's interior have now made a plausible case for bacterial domination in biomass as well.

As Ariel, in *The Tempest,* proclaimed his ubiquity in all manifestations of life—"where the bee sucks, there suck I / In a cowslip's bell I lie"—so, in this world, do bacteria dwell in virtually every spot that can sustain any form of life. And we have underestimated their global number because we, as members of a kingdom far more restricted in potential habitation, never appreciated the full range of places that might be searched.

For example, the ubiquity and role of bacteria in the open oceans have been documented only in the past twenty years. Conventional methods of analysis missed up to 99 percent of these organisms (Fuhrman, McCallum, and Davis, 1992) because we could identify only what could be cultured from a water sample—and most species don't grow on most culture media. Now, with methods of genomic sequencing and other techniques, we can assess taxonomic diversity without growing a large, pure culture of each species.

Scientists had long known that the photosynthesizing Cyanobacteria ("blue-green algae" of older terminology) played a prominent role in the oceanic plankton, but the great abundance of heterotrophic bacteria (non-photosynthesizers that ingest nutrients from external sources) had not been appreciated. In coastal waters, these heterotrophs constitute from 5 to 20 percent of microbial biomass and can consume an amount of carbon equal to 20 to 60 percent of total "primary production" (that is, organic material made by photosynthesis)—giving them a major role near the base of oceanic food chains. But Jed A. Fuhrman and his colleagues then studied the biomass of heterotrophic bacteria in open oceans (that is, needless to say, by far the largest habitat on earth by area) and found that they dominate in these environments. In the Sargasso Sea, for example (Fuhrman et al., 1989), heterotrophic bacteria contribute 70 to 80 percent of microbial carbon and nitrogen, and form more than 90 percent of biological surface area.

When I visited Jed Fuhrman's lab at the University of Southern California, I asked him if he could estimate the earth's total bacterial biomass relative to contributions from the other kingdoms of life. These "back of the envelope" calculations have a long and honorable history in biological barroom discussions—and no one would want to grant them any more technical or firmer status. They must, of necessity, be based on a large number of assumptions and "best estimates" that may be wildly wrong for lack of better available data (average number of bacteria per milliliter of sea water for all the world's oceans, for example). Still, such calculations serve a useful function in defining ballparks. Fuhrman made his best estimate for me, and came up with an oceanic bacterial biomass equal to about one-fiftieth of the entire terrestrial biota, including wood. This may not sound impressive, but whenever such a calculation gets you within an "order of magnitude" or two of a key number, then you are "in the same ballpark." (An order of magnitude—the standard measure of comparison for such rough calculations—is a multiple of ten. Thus, 1/50 is between one [1/10] and two [1/100] orders of magnitude from the terrestrial figure—and definitely in the same ballpark.) This figure is even more impressive when you realize (1) that all traditional estimates have granted domination to the multicells by orders of magnitude because the biomass of wood must be so high; (2) that Fuhrman has not included terrestrial bacteria of soil, gut floras, nodules of leguminous plants, etc.; and (3) that an even greater potential source of biomass from a "new" environment—the earth's interior—has been similarly excluded. If we then turn to some stunning, and controversial, data on the earth's interior, we may really be in for a surprise.

I shall present this new information by snippets in chronological order—a good way to mark successive claims for "internal" bacteria: first around deep sea vents, then in oil reservoirs, and finally in ordinary interior rocks, a finding that, at one extreme of interpretation, makes our surficial biota puny and exceptional, and suggests that interior bacterial biotas may be life's standard and universal mode.

In the late 1970s, marine biologists discovered the bacterial basis of food chains for deep-sea vent faunas—and the unique dependence of this community upon energy from the earth's interior, rather than from a solar source (as discussed on page 185). Two kinds of vents had been described: cracks and small fissures with warm water emerging at temperatures of 40° to 70°F; and large conical sulfide mounds, up to thirty feet in height, and spouting superheated waters at temperatures that can exceed 600°F. Bacteria had been identified in waters from small fissures of the first category, but, unsurprisingly, they "had previously not been thought to exist in the superheated waters associated with sulfide chimneys" (Baross et al., 1982, page 366).

But, in the early 1980s, John Baross and his colleagues discovered a bacterial biota, including both oxidative and anaerobic species, in superheated waters emanating from the sulfide mounds (also known as "smokers"). They cultured bacteria from waters collected at 650°F and then grew vigorous communities in a laboratory chamber with waters heated to 480°F at a pressure of 265 atmospheres. Thus, bacteria can (and do) live in high temperatures (and pressures) of waters flowing beneath the earth's surface (Baross et al., 1982; Baross and Deming, 1983).

Writing about this work in a commentary for *Nature,* Britain's leading journal of professional science, A. E. Walsby (1983) commented, "I must admit that my first reaction on reading the manuscript of Baross and Deming, arriving as it did on the eve of April Fool's day, was one of incredulity." Walsby began his comment by noting that these deep-sea bacteria grow at a heat exceeding the title of Ray Bradbury's famous story, *Fahrenheit 451*—the temperature at which paper ignites (and thought can therefore be more easily controlled by destruction of radical literature). Pressure is the key to an otherwise paradoxical situation. Life needs liquidity, not necessarily coolness. At the enormous pressures of the sea floor, water does not boil at temperatures tolerated by these bacteria. Baross and Deming end their article, prophetically as we shall see, by noting (1983, page 425):

> These results substantiate the hypothesis that microbial growth is limited not by temperature but by the existence of liquid water, assuming that all other conditions necessary for life are provided. This greatly increases the number of environments and conditions both on Earth and elsewhere in the Universe where life can exist.

Then, in the early 1990s, several groups of scientists found and cultured bacteria from oil drillings and other environments beneath oceans and continents—thus indicating that bacteria may live generally in the earth's interior, and not only in limited areas where superheated waters emerge at the surface: from four oil reservoirs nearly two miles below the bed of the North Sea and below the permafrost surface of Alaska's North Slope (Stetter et al., 1993); from a Swedish borehole nearly four miles deep (Szewzyk et al., 1994); and from four wells about a mile deep in France's East Paris Basin (L'Haridon et al., 1995). Water migrates extensively through cracks and joints in subsurface rocks, and even through pore spaces between grains of sediments themselves (an important property of rocks, known as "porosity" and vital to the oil industry as a natural mechanism for concentrating underground liquids—and, as it now appears, bacteria as well). Thus, although such data do not indicate global pervasiveness or interconnectivity of subsurface bacterial biotas, we certainly must entertain the proposition that much of the earth deep beneath our feet teems with microbial life.

The most obvious and serious caution in these data emerges from another general property of bacteria: their almost ineradicable ubiquity. How do we know that these bacteria, cultured from waters collected at depth, really live in these underground environments? Perhaps they were introduced into deeper waters by the machinery used to dig the oil wells and boreholes that provided sites for sampling; perhaps (with even more trepidation) they just represent contamination from ubiquitous and ordinary bacteria of our surface environments, stubbornly living in laboratories despite all attempts to carry out experiments in sterile conditions. (A fascinating, and very long, book could be written about remarkable claims for bacteria in odd places—on meteorites, living in geological dormancy within 400-million-year-old salt deposits—that turned out to be ordinary surface contaminants. I well remember the first "proven" extraterrestrial life on meteorites, later exposed as ordinary ragweed pollen. Ah-choo!)

This well-known possibility sends shivers down the spine of any scientist working in this area. I am no expert and cannot make any general statement. I would not doubt (and neither do the authors of these articles) that some reports may be based on contamination. But all known and possible precautions have been taken, and best procedures for assuring sterility have been followed. Most persuasively, many of the bacteria isolated from these deep environments are anaerobic hyperthermophiles (jargon for bacteria growing at very high temperatures in the absence of oxygen) that could thrive in subterranean conditions, and cannot be laboratory contaminants because they die in ordinary surface environments of "low" temperature and pressure and abundant oxygen.

Writing in *The New York Times* on December 28, 1993, William J. Broad summarized the case nicely:

> Some scientists say the microbes may be ubiquitous throughout the upper few miles of the Earth's crust, inhabiting fluid-filled pores, cracks, and interstices of rocks while living off the Earth's interior heat and chemicals. Their main habitats would be in the hot aquifers beneath the continents and in oceanic abysses, fed perpetually by the nutrients carried by the slow circulation of fluids like oil and deep ground water.

We might ask one further question that would clinch the case for underground ubiquity: Moving away from the specialized environments of deep-sea vents and oil reservoirs, do bacteria also live more generally in ordinary rocks and sediments (provided that some water seeps through joints and pore spaces)? New data from the mid-1990s seem to answer this most general question in the affirmative as well.

R. J. Parkes et al. (1994) found abundant bacteria in ordinary sediments of five Pacific Ocean sites at depths up to 1,800 feet. Meanwhile, the United States Department of Energy, under the leadership of Frank J. Wobber, had been digging deep wells to monitor contamination of groundwater from both inorganic and potentially microbial sources (done largely to learn if bacteria might affect the storage of nuclear wastes in deep repositories!). Wobber's group, taking special pains to avoid the risk of contamination from surface bacteria introduced into the holes, found bacterial populations in at least six sites, including a boring in Virginia at 9,180 feet under the ground!

William J. Broad wrote another article for the *Times* (October 4, 1994), this time even more excited, and justifiably so:

> Fiction writers have fantasized about it. Prominent scientists have theorized about it. Experimentalists have delved into it. Skeptics have ridiculed it. But for decades,

> nobody has had substantial evidence one way or another on the question of whether the depths of the rocky earth harbor anything that could be considered part of the spectacle of life—until now. . . . Swarms of microbial life thrive deep within the planet.

Stevens and McKinley (1995) then described rich bacterial communities living more than three thousand feet below the earth's surface in rocks of the Columbia River Basalt in the northwestern United States. These bacteria are anaerobic and seem to get energy from hydrogen produced in a reaction between minerals in the basaltic rocks and groundwater seeping through. Thus, like the biotas of the deep-sea vents, these bacteria live on energy from the earth's interior, entirely independent of the photosynthetic, and ultimately solar, base of all conventional ecosystems. To confirm their findings in the field, Stevens and McKinley mixed crushed basalt with water free from dissolved oxygen. This mixture did generate hydrogen. They then sealed basalt together with groundwaters containing the deep bacteria. In these laboratory conditions, simulating the natural situation at depth, the bacteria thrived for up to a year.

Following a scientific tradition for constructing humorous and memorable acronyms, Stevens and McKinley have named these deep bacterial floras, independent of solar energy, and cut off from contact with surficial communities, SLiME (for subsurface lithoautotrophic microbial ecosystem—the second word is just a fancy way of saying "getting energy from rocks alone"). Jocelyn Kaiser (1995), writing a comment for *Science* magazine on the work of Stevens and McKinley, used a provocative title: "Can deep bacteria live on nothing but rocks and water?" The answer seems to be yes.

My colleague Tom Gold of Cornell University may be one of America's most iconoclastic scientists. (One prominent biologist, who shall remain nameless, once said to me that Gold ought to be buried deep within the earth along with all his putative bacteria.) But no one sells him short or refuses to take him seriously—for he has been right far too often (we only threaten to bury alive the people we fear).

In a remarkable article entitled "The deep, hot biosphere" and published in the prestigious *Proceedings of the National Academy of Sciences* in

1992, Gold set out the full case (truly universal, or at least potentially so) for the importance of bacterial biotas deep within the earth. (He did this, characteristically, a few years before firm data existed for rich bacterial communities in ordinary subsurface rocks. But he was right again, in this factual claim at least, if not necessarily in all his implications. Gold began his case by asking, "Are the ocean vents the sole representatives of this [deep bacterial life], or do they merely represent the examples that were discovered first?")

Of all living things that might expand the range of life beyond conventional habitats of land and oceans, bacteria are the obvious candidates. They are small enough to fit nearly anywhere, and their environmental range vastly exceeds that of all other organisms. Gold writes: "Of all the forms of life that we now know, bacteria appear to represent the one that can most readily utilize energy from a great variety of chemical sources."

Gold then makes a key estimate—for my argument about domination of the modal bacter, at least—of possible bacterial biomass, given the vast expansion of range into rocks and fluids of the earth's interior. Gold's effort is, of course, another back-of-the-envelope calculation, and must be treated with all the caution always accorded to this genre (but remember that the estimates may also be too low, rather than inflated). A large number of assumptions must be made: How deep do bacteria live? At what temperatures? How much of rock volume consists of pore space where bacteria may live in percolating waters? How many bacteria can these waters hold? Since we do not know the actual values for any of these key factors, we must make a "most reasonable" estimate. If actual values differ greatly from the estimate (as they may well do), then the final figure may be very far wrong. (I trust that nonscientific readers will now grasp why, in this enterprise, we are satisfied with "ballpark" estimates that might be "off" by even an order of magnitude or two.)

In any case, Gold based his number for total bacterial biomass on reasonable, even fairly conservative, estimates for key factors—so if most rocks permeable by water do contain bacteria, then his figure is probably in the right ballpark. Gold assumes an upper temperature range of 230° to 300°F and a depth limit of three to six miles. (If bacteria actually live deeper, their biomass might be much higher.) He calculates the mass of water available for bacterial life by assuming that about 3 percent of rock volume consists of pore spaces. Finally, he estimates that bacterial mass might equal about 1 percent of the total mass of available underground water.

Putting all these estimates together, Gold calculates a potential mass of underground bacteria at 2×10^{14} tons. This figure, he writes, is equivalent to a layer five feet thick spread out over the earth's entire land surface—an amount of biomass, Gold states, that would "indeed be more than the existing surface flora and fauna." As a cautious conclusion to his calculation of underground bacterial biomass, Gold writes:

> We do not know at present how to make a realistic estimate of the subterranean mass of material now living, but all that can be said is that one must consider it possible that it is comparable to all the living mass at the surface.

When one considers how deeply entrenched has been the dogma that most earthly biomass lies in the wood of our forest trees, this potentially greater weight of underground bacteria represents a major revision of conventional biology—and quite a boost for the modal bacter. Not only does the earth contain more bacterial organisms than all others combined (scarcely surprising, given their minimal size and mass); not only do bacteria live in more places and work in a greater variety of metabolic ways; not only did bacteria alone constitute the first half of life's history, with no slackening in diversity thereafter; but also, and most surprisingly, total bacterial biomass (even at such minimal weight per cell) may exceed all the rest of life combined, even forest trees, once we include the subterranean populations as well. Need any more be said in making a case for the modal bacter as life's constant center of maximal influence and importance?

But Gold does take one further, and equally striking, step. We are now fairly certain that ordinary life exists nowhere else in our solar system— for no other planetary surface maintains appropriate conditions of temperature and liquid water. Moreover, such earthly surface conditions are probably rare in the universe, making life an unusual cosmic phenomenon.

But the environment of the earth's shallow interior—liquid flowing through cracks and pore spaces in rocks—may be quite common on other worlds, both in our solar system and elsewhere (frozen surfaces of distant planets will not permit life, but interior heat may produce liquid—and a possible environment for life at bacterial grade—within underground rocks). In fact, Gold estimates that "there are at least ten other planetary bodies [including several moons of the giant planets] in our solar system that would have had a similar chance for originating microbial life" because "the circumstances in the interior of most of the solid planetary bodies will not be too different from those at a depth of a few kilometers in the Earth."

Finally, we may need to make a complete reversal of our usual perspective and consider the possibility that our conventional surface life, based on photosynthesis, might be a very peculiar, even bizarre, manifestation of a common universal phenomenon usually expressed by life at bacterial grade in the shallow interior of planetary bodies. Considering that we didn't even know only ten years ago such interior life existed, the transition from unknown to potentially universal must be the most astonishing promotion in the history of favorable revisions! Gold concludes:

> The surface life on the Earth, based on photosynthesis, for its overall energy supply, may be just one strange branch of life, an adaptation specific to a planet that happened to have such favorable circumstances on its surface as would occur only very rarely: a favorable atmosphere, a suitable distance from an illuminating star, a mix of water and rock surface, etc. The deep, chemically supplied life, however, may be very common in the universe.

The modal bacter, in other words, may not only dominate, even by weight, on earth, but may also represent life's only common mode throughout the universe.

No Driving to the Right Tail

A proper theory of morality depends upon the separation of intentions from results. Tragic deaths may occur as unintended consequences of decent acts—and we rightly despise the cold-blooded killer, while holding sympathy for the good Samaritan, even if an unnecessary death becomes the common result of such radically different intentions (the robber who shoots the store owner, and the policeman who kills the same owner because he fired at the robber and missed).

Similarly, any proper theory of explanation in natural history depends upon the distinction of causes and consequences. Darwin's central theory holds that natural selection acts to increase adaptation to changing local environments. Therefore, features built directly by natural selection—the thick coat of the woolly mammoth in my example on page 139, for example—evolve for adaptive reasons by definite cause. But many features that become vital to the lives of their bearers may arise as uncaused (or at least indirectly produced) and "unintended" sequelae or side consequences. For example, our ability to read and write has acted as a prime mover of contemporary culture. But no one could argue that natural selection acted to enlarge our brains for this purpose—for *Homo sapiens* evolved brains of modern size and design tens of thousands of years before anyone thought about reading or writing. Selection made our brains large for other reasons, while reading and writing arose later as a fortuitous or unintended result of an enlarged mental power directly evolved for different functions.

Our intuitions tell us—quite rightly in this case, I believe—that this distinction between *results directly caused* and *consequences incidentally arising* is both important in explaining any particular feature of the organic world and fundamental to any general understanding of evolution. The main issue is not predictability—for a phenomenon may be predictable whether it arises directly for cause or incidentally as a consequence. The key question centers on the nature and character of explanation. The purposeful killer and the erring policeman produce the same result (and with equal predictability in the old-fashioned Newtonian sense of potential for deducing the outcome once we know the positions of all people, the sight line of the gun, the timings, etc.)—yet we yearn to judge the meaning differently based on the distinction between intention and accident.

In the same way, a right tail of increasing maximal complexity might arise on the bell curve of life either (as tradition has held) because evolution inherently drives life to higher levels of complexity or (as I argue in the major claim of this book) as an incidental side consequence of life's necessary origin at the left wall of minimal complexity followed by successful expansion thereafter with retention of an unvarying bacterial mode. Our intuitions detect a radical difference in meaning between these two pathways to predictable production of the same result—and our intuitions are right again. We do, and should, care profoundly about the different meanings—for, in one case, increasing complexity is the driving *raison d'être* of life's history; while, in the other, the expanding right tail is a passive consequence of evolutionary principles with radically different main results. In one case, progress rules and shapes the history of life as the central product of fundamental causes; in the other, progress is secondary, rare, incidental, and shaped by no cause working directly in its interest.

This issue of directly caused results versus incidental consequences has reverberated throughout the history of evolutionary thought. A large literature, both scientific and philosophical, has been devoted to explicating these vital distinctions. A daunting and somewhat jargony terminology has arisen (some, I confess, of my own construction) to carry this debate in the technical literature—adaptations versus exaptations, aptations versus spandrels, selection versus sorting (see Sober, 1984; Gould and Lewontin, 1979; Gould and Vrba, 1982; Vrba and Eldredge, 1984). We will stick to the vernacular here, and make our main distinction between intended results and incidental consequences.

As the main claim of this book, I do not deny the phenomenon of increased complexity in life's history—but I subject this conclusion to two restrictions that undermine its traditional hegemony as evolution's defining feature. First, the phenomenon exists only in the pitifully limited and restricted sense of a few species extending the small right tail of a bell curve with an ever-constant mode at bacterial complexity—and not as a pervasive feature in the history of most lineages. Second, this restricted phenomenon arises as an incidental consequence—an "effect," in the terminology of Williams (1966) and Vrba (1980), rather than an intended result—of causes that include no mechanism for progress or increasing complexity in their main actions.

At most, one might advance Thomas's (1993) claim that "progressive emergence of increasing complexity over the long term is the main effect of evolution. As such, it compels our attention." In other words, Thomas admits that increasing complexity is an incidental consequence, an effect

rather than a main result of causes framed in its interest. He holds, however, that progress still compels our attention as the "main" effect among
all of evolution's incidental consequences. But what possible criterion can
validate this claim beyond the parochial and subjective desire to designate
as primary an effect that both led to human life and placed us atop a heap
of our own definition? I think that any truly dominant bacterium would
laugh with scorn at this apotheosis for such a small tail so far from the
modal center of life's main weight and continuity. I do realize that bacteria can't laugh (or cogitate)—and that philosophical claims for our greater
importance can be based on the consequences of this difference between
them and us. But do remember that we can't live on basalt and water six
miles under the earth's surface, form the core of novel ecosystems based
on the earth's interior heat rather than solar energy, or serve as a possible
model for cosmic life in most solar systems.

In other words, progress as a purely incidental consequence (and limited to a small right tail) just won't do as a validation for our traditional
hopes about intrinsic human importance—the spin-doctoring that pre-
vents the completion of Darwin's revolution in Freud's crucial sense of
pedestal smashing (see chapter 2). I think that virtually every evolutionist
who has ever considered the issue in the terms of this book (that is, as a
history of variation in all life—the full house—rather than as a tale told
by abstracted means or extreme values only) has come to the conclusion
that the appearance of progress as an expanding right tail must arise as an
incidental consequence, not as a main result.

The traditional hope for intrinsic progress as an explicit result must
therefore rest upon a fallback position—not nearly so grand as the original formulation, but a source of some potential solace nonetheless. Even
if we must admit that an expanding right tail arises as an incidental consequence of origin at a left wall with subsequent proliferation, could we
not also hold that other forces operate as well on life's bell curve—and that
some of these other forces do include an intrinsic and predictable drive to
progress?

As stated in point 6 of my epitome (see page 173), such an argument
could be true, would take the following form, and can be tested empirically: life as a whole begins at the left wall and is therefore free to expand
in only one direction. Therefore we cannot use life-as-a-whole to test for

drives to progress—because upward movement of the mean must, in part, reflect the left wall's constraint, not any potential drive. But if we could study the history of smaller lineages with founding members far from the wall—and therefore free to vary in either direction—then we could devise a clear test for general progress. Do such "free" lineages show a tendency for increases in complexity to be more frequent, or greater in effect, than decreases? If most free lineages show a trend to increasing complexity, then we could assert a general principle of progress as a main result for its own sake. The full phenomenon of life's expanding right tail would then arise by two separate and reinforcing processes: an incidental consequence based on constraints of origin at the left wall, and a direct result of intrinsic bias to greater complexity in lineages free to vary in both directions.

This conjecture is logically sound but, by all evidence so far in hand, empirically wrong. I would raise two arguments against intrinsic progress, the first briefly and subjectively, the second at greater length and based upon some compelling recent evidence.

First, if I were a betting man, I would wager a decent sum (but not the whole farm) on a small natural preference for *decreasing* complexity within lineages, and not for the traditional increase, if any general bias exists at all. I make this surprising claim because natural selection, in its purest form, only yields adaptation to changing local environments. These changes should be effectively random (with respect to "progress"), for fluctuations in climate show no temporal trend. A bias for or against increasing complexity therefore requires a general advantage for one direction as life plays its Darwinian game. I can think of a reason why a bias for decreasing complexity might exist, but I cannot defend any corresponding preference for increases. Hence I would bet that a slight overall bias for decreasing complexity might well prevail in the aggregate of all lineages.

I have long been entirely underwhelmed by the standard arguments for general advantages of increasing complexity in the Darwinian game—adaptive benefit of more elaborate bodily form in competition for limited resources, for example. Why should more complex conformations generally prevail? I can imagine such an argument for mammalian brains—if complexity translates to rising flexibility and computing power. But I can envisage just as many situations where more elaborate forms might be a hindrance—more parts to fail, less flexibility because all parts must interact with precision.

But one common mode of Darwinian success (local adaptation) does entail an apparent preference for substantial decreases in complexity—namely, the lifestyle of parasites. We are not speaking here of an organic rarity, but of a mode of life evolved by probably hundreds of thousands of species—a substantial percentage of all living forms. Not all parasites gain adaptive benefit through simplification, but one large group of species certainly does—those that live deep within the bodies of their hosts, permanently attached and receiving all their nutrition by commandeering the blood supply, or some of the food already digested by the host. Such species require neither organs of locomotion nor digestion, and natural selection favors their loss. One or a few novel organs might evolve for special needs—hooks for attaching to the host, or suction devices to drain off food, for example—but these elaborations are more than offset by a far greater number of lost organs.

Often these immobile parasites become little more than bags or tubes of reproductive tissue—simple machines for propagation attached to the internal organs of their host. *Sacculina,* the famous barnacle parasite of crabs and other crustaceans, consists of a formless sac (acting as a brood pouch) attached to the crab's abdomen, with a stalk protruding inside to a system of roots that drain food from the crab's blood spaces. A twenty-foot-long tapeworm in a human intestine may contain of hundreds of sections (strobilae), each little more than a simple sac containing members of the next generation. The entire phylum Pentastomida, parasites of the respiratory tract of vertebrates, builds an elaborate organ for sucking blood, but no internal parts for locomotion, respiration, circulation, or excretion.

Thus, if "standard" natural selection on free-living creatures produces no bias in either direction, and if parasites tend to become simplified while no countervailing bias toward greater complexity exists, then a small overall tendency toward decreasing complexity may characterize the history of most lineages (as their parasitic species simplify, while their free-living species show no trend). Please note that the right tail for the full bell curve of life will still expand through time—even if a bias toward decreasing complexity operates in most lineages. For species moving left to less complexity enter a domain already inhabited, while rarer species moving right may enter a previously unoccupied realm of complexity. The drunkard will end up in the road even if, for some reason, he moves more often toward the wall than toward the gutter—for he bounces off the wall but falls prostrate (and permanently) in the gutter. An entire system can extend its extreme in one direction even if individual lineages have a bias for excursions in the other direction.

But I can also think of an argument against my own claim for parasites. Adult forms do indeed tend to evolve toward greater simplicity, but when we confine our attention to adults, we fall into another conventional bias (not as general or pervasive, no doubt, as our preferences for progress, but a seriously distorting limitation nonetheless). A human being is not defined by the nongrowing form of adult years; kids are people too. Evolution shapes a full life cycle, not only an adult body. The immobile blood-sucking or food-draining adult parasite may have evolved toward greater simplicity compared with free-living ancestors, but full parasitic life cycles often change in the other direction toward great elaboration, sometimes with adaptation to two or three different hosts in the course of a full ontogeny.

The adult *Sacculina* may be an external blob attached to some internal roots, but the larval life cycle is astonishingly complex (see Gould, 1996)—several free-living planktonic forms, followed by a settling phase that cements to the crab, grows a dart that pierces the crab's body, and then injects the few cells that eventually grow into the adult blob and roots. Similarly, pentastome larvae first bore through the gut of an initial host. When a vertebrate eats its first home, the matured pentastome moves to the respiratory tract either by crawling from the vertebrate's stomach to the esophagus and then boring through, or by tunneling through the intestinal wall and into the bloodstream. The pentastome then attaches to its final site by means of complex hooks surrounding the mouth.

I therefore have little confidence that we can specify a clear bias one way or the other on general principles. But we do have a wealth of empirical data available for study. After all, the founding species of most multicellular lineages does not begin at a wall—and subsequent evolution remains free to produce either more or less complex species. If we can agree on a measure of complexity, and document enough lineages, we may be able to extract a general conclusion. This subject has just begun to interest paleontologists in the past few years. We have not yet compiled nearly enough cases for any confident general solution. But the initial studies offer great promise, for we have at least made this vital subject tractable and testable. And the first few cases all point in the same radical direction— no bias toward increasing complexity has yet been measured.

This line of research has been pioneered by Dan McShea of the University of Michigan, now at Santa Fe's Institute for the Study of Complexity. Much of the technical literature must focus on providing an unambiguous and quantifiable definition for a very fuzzy vernacular term with a wide variety of meanings, some contradictory—namely, complexity itself. What do we mean when we say that a thing is more complex than something else? Several criteria fit our vernacular sense, depending upon the context. Complexity has morphological, developmental, and functional aspects. A junk heap (to use an example favored by McShea and Thomas) may be morphologically very complex (in consisting of so many highly varied and independent parts) but functionally quite simple (just glop for a landfill). On the other hand, what is functionally simple for us might be quite complex to other users—in this case, to the seagull who must distinguish all the little bits while searching for morsels of food.

I do not wish to address this technical subject at length in a book for general readers (but see McShea, 1992, 1993, 1994, and Thomas, 1993, for interesting discussion), though the importance and nature of the problem must be recorded. I do not think that any general solution can be found— because "complexity" is a vernacular term with several legitimately different meanings, and we may well be interested in all of them. For science as "the art of the soluble" (to use P. B. Medawar's felicitous phrase)—an enterprise dedicated to posing answerable questions—we must only resolve that we will choose a rigorously quantifiable definition of complexity and be very clear about which aspects of vernacular meaning will be thus addressed, and which omitted. (Someone else, or you yourself in a subsequent study, may then measure other aspects of complexity.) The literature has been admirable on this account, and therefore happily free of the muddiness that accompanies so much science.

McShea has favored a morphological definition—not because he views this meaning as closer to some vernacular norm, but because it permits well-defined measurement and rigorous testing. He writes (1996): "The point is to rescue the study of biological complexity from a swamp of impressionistic evaluations, biased samples, and theoretical speculations, and to try to place it on more solid empirical ground." McShea employs the following conceptualization to construct his quantifications:

> The complexity of a system is generally acknowledged to be some function of the number of different parts it has, and of the irregularity of their arrangement. Thus, heterogenous, messy, or irregularly configured systems are complex, such as organisms, automobiles, compost heaps, and junk yards. Order is the opposite of complexity. Ordered systems are homogeneous, redundant, or regular, like picket fences and brick walls (1993, page 731).

In his major study of the vertebrate spinal column, for example, Mc-Shea (1993) operationalizes this definition by measuring complexity as the

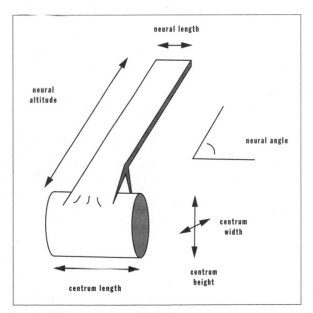

FIGURE 32
Measurements used by McShea to assess the history of complexity in vertebrae.

degree to which individual vertebrae differ among themselves. (In the less complex backbone of a fish, forty or more vertebrae may be effectively alike as simple discs of similar size; the more complex mammalian spine has fewer vertebrae differentiated into the varied forms and sizes of neck bones, back vertebrae, and sacral discs that support the pelvis.) In practice (see Figure 32), McShea measures six variables (five linear dimensions and an angle) on each vertebra and then calculates the difference among vertebrae. He uses three assessments of complexity as variation among vertebrae: (1) the maximal difference between any two vertebrae of the same spinal column; (2) the average difference between each vertebra and the mean for all vertebrae; and (3) the average difference between each pair of adjacent vertebrae.

McShea's framework for testing harmonizes perfectly with the perspective of this book. He holds that trends come in two basic modes with strikingly different fundamental causes. He names these categories *driven* and *passive,* and argues that they represent natural "kinds," not just conceptual conveniences for human understanding. He writes (1994, page 1762): "These results do raise the possibility that the passive and driven mechanisms may be natural categories and that they may correspond to distinct and well-defined causes of large-scale trends."

Driven trends correspond to the traditional view of an overall movement achieved because each element evolves with a bias for change in this direction. A driven trend to complexity would arise because evolution generally favors more complex creatures—and each species of a lineage therefore tends to change in this manner. (In other words, natural selection acts as a driver, conveying each vehicle in a favored direction.) Passive trends (see Figure 33) conform to the unfamiliar model, championed for complexity in this book, of overall results arising as incidental consequences, with no favored direction for individual species. (McShea calls such a trend passive because no driver conducts any species along a preferred pathway. The general trend will arise even when the evolution of each individual species confirms to a "drunkard's walk" of random motion.) For passive trends in complexity, McShea proposes the same set of constraints that I have advocated throughout this book: ancestral beginnings at a left wall of minimal complexity, with only one direction open to novelty in subsequent evolution.

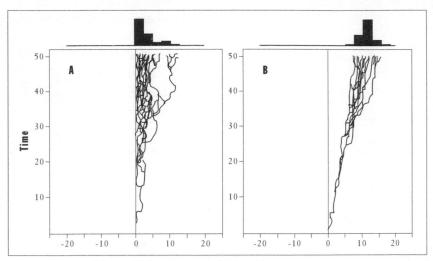

FIGURE 33

Passive and driven trends in McShea's terminology. A passive trend *(A)* begins near a left wall, retains a constant mode at this beginning position, and expands in the only open direction toward the right. In a driven trend *(B)*, both minimum and maximum values increase through time.

McShea proposes three tests for distinguishing driven from passive trends:

1. THE TEST OF THE MINIMUM.   In passive systems, minimum values of complexity should be preserved by some species throughout the expanding history of a lineage because no general evolutionary preference for complexity exists, and some species should therefore do best by remaining simple. In driven systems, both minimum and maximum complexity should increase through time because higher complexity confers such general advantages that evolution of all species should be biased in this direction. (The preservation and continuing enhancement of life's bacterial mode strongly points to the passive mode for life as a whole.)

This test, although indicative, does not fully distinguish passive from driven trends because even a driven trend might permit a few species to retain minimal values. (In a driven trend, the minimum might not disappear, but these lowest values should at least become less frequent over time.)

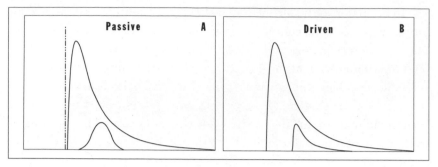

FIGURE 34

A test to distinguish passive from driven trends. The entire distribution for the passive
trend should be right skewed, but component lineages that begin far away from the left wall
should have normal distributions.

2. THE TEST OF ANCESTOR-DESCENDANT PAIRINGS.   This powerful and
obvious test identifies an ancestral species for an expanding lineage and
then simply tabulates all descendants to judge whether most become more
complex, simpler, or stay the same. In principle this is the most decisive
test of all. But in practice we cannot always use it because the fossil record
is so imperfect. We often do not know the ancestral species, or we do not
have enough descendants to make a proper randomized test of subsequent directions.

3. THE TEST OF SKEWING.   For life-as-a-whole, both the passive and
driven mechanism can produce the same overall result of a right-skewed
distribution with an expanding tail at maximal complexity. McShea argues that we might distinguish passive from driven modes by studying the
skewness of component lineages that begin far from the wall and can
therefore vary in either direction (see Figure 34). In driven systems, the
component lineages should also tend to be right skewed because all species
experience the bias of progress as a favored direction and should therefore
contain more species moving along this preferred pathway, thereby
stretching the entire distribution toward the right. But in passive systems,
component lineages should develop no skew because increases and decreases in individual species should be equally common—that is, as many
species should move leftward to less complexity as rightward to more
elaboration.

In his major study, McShea (1993, 1994) has applied these tests to evolution of the vertebral column. A general trend obviously exists for vertebrates as a whole because the first vertebrates were fishes with a backbone built of essentially identical elements, while later mammals evolved considerable variation among vertebrae along the spinal column. But is this trend passive or driven? (Tradition says driven, but one fact certainly leaves maximal "room" for passivity. Much like the initial living thing at the left wall of minimal complexity, or the founding foraminiferal species at the absolute left wall of minimal sieve size [see pages 157–158], vertebrates begin at a theoretically minimal value of complexity by McShea's measurements. Since the founding fishes tend to have vertebral columns made of identical elements, their measured complexity will be close to flat zero [McShea measures complexity as differences among vertebrae]. There really is no place to go from this initial point but up!)

McShea's study of sublineages within mammals provides strong quantitative evidence for increasing general complexity in the passive mode— thus supporting the claim of this book that no explicit preference or bias for complexity acts as a driving force in the evolution of life. McShea surveyed five sublineages where he could identify or infer an ancestor, thus permitting him to use the most powerful second test: ruminant mammals (the large group of cud-chewing cattle, deer, etc.); squirrels of the large family Sciuridae; the entire order of pangolins (a group of armored anteaters, now represented by the genus *Manis* of Africa and Asia); whales; and camels.

All the tests provide evidence for a passive trend and no drive to complexity. McShea found twenty-four cases of significant increases or decreases in comparing the range of modern descendants with an ancestor (out of a total potential sample of ninety comparisons, or five groups of mammals, each with six variables measured in each of three ways; for the other comparisons, average descendants did not differ significantly from ancestors). Interestingly, thirteen of these significant changes led to *decreases* in complexity, while only nine showed increase. (The difference between thirteen and nine is not statistically significant, but I am still wryly amused, given all traditional expectation in the other direction, that more comparisons show decreasing rather than increasing complexity.)

McShea was then able to apply the third test of skewing to three vertebral dimensions measured in three lineages. The mean skew value for all nine distributions is actually negative (−.19), not significantly so to be sure, but quite a comeuppance for the traditional view of complexity as driven—a conclusion that implies positive (right) skewing for component lineages!

McShea then summarizes his entire study (1994, page 1761):

> The minimum complexity of vertebral columns probably did not change (indeed, the actual minimum seems to have remained close to the theoretical minimum), ancestor-descendant comparisons in subclades of mammals revealed no branching bias, and the mean subclade skew was negative, all pointing to a passive system.

One study doesn't prove a generality any more than a single swallow makes a summer, but when our first rigorous data point to a conclusion so at variance with traditional views, we must sit up and take notice, and then go out to make more tests. The few other available studies also support the passive rather than the driven mode. In an interesting report, presented at paleontological sessions at the 1995 annual meeting of the Geological Society of America in New Orleans, McShea provided the first results for a quite different meaning of complexity—developmental rather than morphological, and defined as the number of independent growth factors that build a structure through embryology (practically measured as correlation coefficients between pairs of measures, with perfect positive correlation indicating that the two measures represent only one mode of growth, and zero correlation implying that the two measures indicate different developmental influences).

Working with Benedikt Hallgrimsson and Philip D. Gingerich, McShea applied this method to a large series of classical and excellent data on measurements of fossil teeth, compiled over many years by Gingerich on evolutionary sequences for several mammalian lineages in the Bighorn Basin of Wyoming. They found no trend to increasing complexity and concluded: "Tests detected no bias, no tendency for non-hierarchical developmental complexity either to increase or decrease."

In the only other comprehensive study, using an interesting metric for complexity applied to a very different group of organisms, Boyajian and Lutz (1992 and personal communication) studied one of the classic examples of supposedly driven evolution toward greater complexity—and again found evidence only for the passive mode!

Ammonites are extinct relatives of the modern chambered nautilus—coiled cephalopod shells housing animals related to modern squid and octopuses. Internal chambers meet the external shell at a boundary known as a "suture line." In nautiloids the suture line is usually straight or mildly wavy, but in ammonites the suture line can become intricately sinuous and digitated. In the everyday sense that sinuous and digitated looks more complicated than straight or mildly wavy, an old paleontological truism asserts that ammonite sutures become more complex through time. Ever since the earliest days of paleontology, increasing complexity of the ammonite suture has ranked among the two or three "classic" trends that "everybody knows" in the fossil record of invertebrates.

Boyajian and Lutz used a clever measure of "fractal dimension" to assess the complexity of ammonite sutures. (Heretofore, the trend has merely been asserted subjectively, rather than proven quantitatively, because no one could figure out a rigorous way to measure the complexity of such a twisty line.) Fractals have become a hot topic of popular culture, but in a technical sense, fractals are curves and surfaces that exist between ordinary dimensions. Since a straight line has a fractal dimension of one, and a plane a fractal dimension of two, twisty lines must measure between one and two—that is, between a minimum of one for the straight line between two points, and the unattainable maximum of two for a line that twists and turns so much that it fills an entire plane between the two points at opposite edges. The higher the fractal dimension, the more "complex" the suture in our visceral and traditional sense that the squiggliest lines are most elaborate. Boyajian and Lutz measured the fractal dimension of a suture in each of 615 genera of ammonites spanning the full range of their history. The measured scope of fractal dimensions runs from just a tad over 1.0 (very simple sutures are close to straight lines) to just over 1.6 for the most complex.

All early ammonites grew fairly simple sutures, and some measure near the theoretical minimum of 1.0 for a straight line. Thus, as in Mc-Shea's vertebrae, where founders showed minimal complexity, any move-

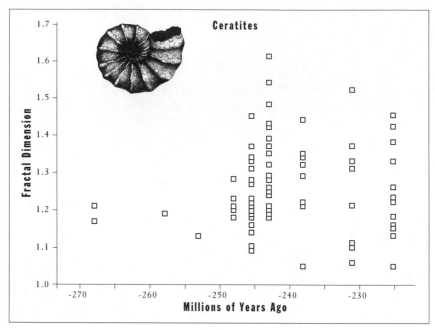

Evolution of complexity as measured by fractal dimension in a group of ammonites (the cer-
atites). The first species (left) have simple sutures near the left wall. Low values persist and
even decrease through the group's history, but variation also expands into the only open direction of higher fractal dimension.

ment away from initial values could only be upward! This origin at a true
left wall set the supposedly driven trend to increasing sutural complexity—
for many later ammonites have very complex sutures, and scientific imagination can always drum up some putative adaptive reason for why
complex sutures should be better, and therefore favored by natural selection (greater shell strength against hydrostatic pressure, and increasing area
for attachment of muscles, have been favored).

But Boyajian and Lutz could find no evidence for a driven trend; all
data identify the trend as probably passive—an incidental effect of minimally simple beginnings at a left wall, followed by no bias whatever for
increasing complexity in individual lineages thereafter. Most lineages of
ammonoids maintain species of low complexity throughout their history
(see Figure 35 for an example). Most important, Boyajian and Lutz found
no bias for increasing complexity among all the ancestor-descendant pairs

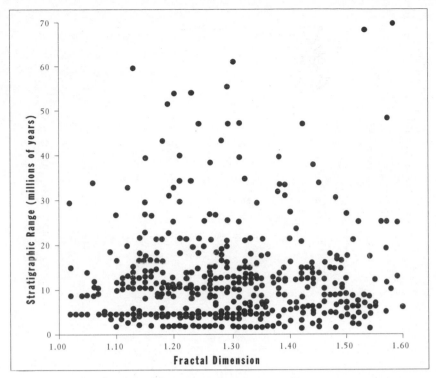

FIGURE 36

Longevity (in millions of years) for ammonite genera on the vertical axis plotted against fractal dimension on the horizontal axis. There is no correlation between complexity and success as measured by longevity.

they could specify (note the similarity to Arnold et al.'s discovery of no bias for increasing size in ancestor-descendant pairs of forams—see page 161). Finally, if complexity is such a good thing, then genera with more complex sutures should live longer. But Boyajian and Lutz found no correlation between sutural complexity and geological longevity (see Figure 36).

Only the most minuscule proportion of scientific studies ever gets reported in the press, and these decisions often bear little correlation with the importance of such studies for professionals. Better relationships can be found between the decision to report and the degree to which a conclusion disturbs conventional notions (often misconceptions) about the nature of things. The research of McShea and Boyajian is important to professionals, but their studies also received rare and extensive coverage in the popular press because they challenged something that "everybody knows"—and that turns out to be probably wrong: the supposed drive to increasing complexity as the defining feature of life's evolutionary thrust. Consider the leads to the two major press accounts. Carol K. Yoon in *The New York Times* (March 30, 1993):

> Surveying life's rich parade from the first single-celled be-
> ings in the primordial soup to the diverse array of organ-
> isms into which they developed, evolutionary biologists
> have marveled at the ever more complex cast of creatures
> that has continued to grace the planet. The evolution of
> larger brains, more efficient metabolisms and more elab-
> orate social systems all seem to support the conventional
> wisdom that complexity increases during evolution. So
> clear is the trend that some biologists suggest that the
> evolutionary process is actually driving the increase in
> complexity.... But in two of the first studies to measure
> these trends, based on mammals' backbones and fossil
> shells, researchers say they have been unable to detect any
> overall evolutionary drive toward greater complexity.

And Lori Oliwenstein in *Discover* (June 1993):

> Everybody knows that organisms get better as they
> evolve. They get more advanced, more modern, and less
> primitive. And everybody knows, according to Dan
> McShea (who has written a paper called "Complexity
> and Evolution: What Everybody Knows"), that organ-
> isms get more complex as they evolve. From the first cell
> that coalesced in the primordial soup to the magnificent
> intricacies of *Homo sapiens,* the evolution of life—as
> everybody knows—has been one long drive toward
> greater complexity. The only trouble with what every-
> body knows ... is that there is no evidence it's true.

Few intellectual tyrannies can be more recalcitrant than the truths that everybody knows and nearly no one can defend with any decent data (for who needs proof of anything so obvious). And few intellectual activities can be more salutary than attempts to find out whether these rocks of ages might crumble at the slightest tap of an informational hammer. I love the wry motto of the Paleontological Society (meant both literally and figuratively, for hammers are the main tool of our trade): *Frango ut patefaciam*—I break in order to reveal.

A Note on the Fatal Weakness of the Last Straw

People under assault, and hopelessly overmatched, often do the opposite of what propriety might suggest: they dig in when they ought to accommodate. We call this behavior "siege mentality." Davy Crockett, Jim Bowie, and Co. won posthumous immortality by their intransigence at the Alamo, but an honorable surrender (given their hopeless situation and the certainty of carnage with continued fighting) might have secured the more worldly privilege of telling good war stories over a beer at a Texan bar (for independence from Mexico would have been won in any case) some twenty years later.

I believe that the power of arguments against inherent progress as a driving force of evolution, and the strength of data on the modal bacter and the passive character of trends to the right tail, must now evoke something like a siege mentality among those who still wish to maintain that evolution validates the primacy and domination of human life on our planet. To what may such people now turn as a source of natural solace? The modal bacter must be acknowledged as dominant by any reasonable criterion. The right tail exists, but only as a little appendage that cannot wag the dog of life's full house. Moreover, the right tail arose as an incidental consequence, in a passive trend powered by constraints of life's origin next to the left wall—and not by any cause or bias that favored increasing complexity as a natural good, and a driving thrust of evolution.

The embattled traditionalist must therefore stand his ground on the right tail of his natural habitat. He must adopt a siege mentality and dig in to protect his own restricted turf. The right tail, he must now admit, may be small and merely consequential. But grant me, he pleads, this last potential natural comfort: "May I not at least be a king in my own restricted castle? I once thought that my domain extended over all nature—that all else must be viewed as predictably preparatory to my eventual origin. I am now prepared to admit the hubris and falsity of this view. I reside on a small and incidental tail. But I am, at least, the creature of maximal complexity ('properly' defined by neural elaboration) on this tail, which I therefore dominate by right. This right tail, however passively fashioned, still had to develop and, ultimately, had to spawn a creature like me. Give me, then, at least, this one remaining solace in a parody of a fine old song: 'It had to be me, wonderful me; it had to be me.'

"Let me, in short, live like Pio Nono (the nineteenth-century Pope Pius IX). My predecessors held temporal power over much of Europe. I once ruled a good part of Italy, though I am now confined to a tiny principality—Vatican City—within Rome. But at least my rule here is absolute—and I can proclaim my infallibility!"

But even this reverie—a bit manic to be sure, for embattlement tends to inspire paranoia and delusions of limited grandeur—cannot be sustained. The claim that a conscious creature like us must evolve because we can predict the development of an expanding right tail for all of life represents a classic "category mistake"—in this case, the false inference of a particular from a valid generality. The right tail did predictably arise (if only as a passive consequence), but any individual creature on the right tail of earthly life at this particular time represents a fortuitous and improbable result, one actualization among a hundred million unrealized alternatives. Wind back the tape of life to the origin of modern multicellular animals in the Cambrian explosion, let the tape play again from this identical starting point, and the replay will populate the earth (and generate a right tail of life) with a radically different set of creatures. The chance that this alternative set will contain anything remotely like a human being must be effectively nil, while the probability of any kind of creature endowed with self-consciousness must also be extremely small.

This theme of radical contingency and improbability for particulars, whatever the predictability of general patterns (with humans clearly defined as an improbable particular, not part of any expected generality), does not fall under the scope of this book. But I do need to epitomize the argument at this point (as abstracted from my previous book *Wonderful Life*) because the traditional view, challenged and upended by contingency, forms the last refuge for a hope that we might validate human supremacy as an expected result of general evolutionary principles.

Under the traditional model of evolutionary history as a "cone of increasing diversity," life moves ever upward to greater progress, and outward to a larger number of species—from simple Cambrian beginnings for multicellular animals to our modern levels of progress and range of diversity. Under this iconography, pathways actually followed run along predictable courses that would be at least roughly repeated in any replay. But a radically different view, suggested by a thorough restudy of soft-bodied fossils in the Burgess Shale and other Cambrian faunas, indicates that an inverted iconography may be more appropriate—with maximal anatomical range of disparity reached early in life's history, followed by the extinction of most initial experiments and the "settling down" of life's diversity to just a few of the original possibilities. Moreover, we have strong reason to suspect that the loss of most, and survival of just a few, occurred more through a distribution of lottery tickets than by victories for predictable cause based on higher levels of progress among winners. In the "pure" lottery model, "tickets" are distributed at random and few initial lineages receive such a blessing. Any replay distributes the tickets to another random set, and leads to a radically different group of survivors. Since our own lineage of vertebrates held a tenuous position among these initial experiments—with only two early Cambrian precursors known as fossils, *Pikaia* from the Burgess Shale, and *Yunnanozoon* recently described from Chengjiang in China (see Chen et al., 1995, and Gould, 1995)—we must assume that most replays would not include the survival and flourishing of vertebrates. All of us—from sharks to rhinos to humans—would then have been excluded from the history of life.

If this good fortune of radical contingency occurred but once, with predictability based on progress prevailing thereafter, then we might view human emergence as close to inevitable following one lucky spin of fortune's wheel. But radical contingency is a fractal principle, prevailing at all scales with great force. At any of a hundred thousand steps in the particular sequence that actually led to modern humans, a tiny and perfectly plausible variation would have produced a different outcome, making history cascade down another pathway that could never have led to *Homo sapiens,* or to any self-conscious creature.

If one small and odd lineage of fishes had not evolved fins capable of bearing weight on land (though evolved for different reasons in lakes and seas), terrestrial vertebrates would never have arisen. If a large extraterrestrial object—the ultimate random bolt from the blue—had not triggered the extinction of dinosaurs 65 million years ago, mammals would still be small creatures, confined to the nooks and crannies of a dinosaur's world, and incapable of evolving the larger size that brains big enough for self-consciousness require. If a small and tenuous population of protohumans had not survived a hundred slings and arrows of outrageous fortune (and potential extinction) on the savannas of Africa, then *Homo sapiens* would never have emerged to spread throughout the globe. We are glorious accidents of an unpredictable process with no drive to complexity, not the expected results of evolutionary principles that yearn to produce a creature capable of understanding the mode of its own necessary construction.

·15·

An Epilog on Human Culture

Most of this chapter has focused on constraints imposed by life's origin at a left wall of minimal complexity, followed by a passive trend to the right as life diversified. As in all other examples for this book, I emphasized how explicit consideration of all the variation (the "full house") can engender proper understanding, while the old Platonic strategy of abstracting the full house as a single figure (an average construed as an archetype, or an extreme example to excite our wonder or horror), and then tracing the pathway of this single figure through time, usually leads to error and confusion.

My two major examples in this book—the extinction of 0.400 hitting in baseball and the absence of a driven trend to complexity in the history of life—consider different sides of the same analytical strategy (studying the full house rather than the abstracted essence). The baseball example speaks of encroachment upon a right wall of human limitations; the history of life invokes expansion away from a left wall of minimal complexity. In this second example, I viewed life as expanding passively into a rightward domain of increasing elaboration—but I never addressed the principle that some constraint might eventually limit the spread by acting as a right wall. The baseball example illustrates the shaping power of right walls at the apogee of human achievement—and we should also consider their potential role in the history of human life.

We live in a world of limits. Goethe, citing an old German proverb, wrote: *Es ist dafür gesorgt, dass die Bäume nicht in den Himmel wachsen* (it is ordained that trees cannot grow to heaven). Such mechanical constraints are easily appreciated (and quantified) for objects of human or natural construction. My native state of New York has adopted a motto of one word: *Excelsior*—or "ever upward." But not all the way to heaven. . . . I once stood before a picture window on the twenty-fifth floor of a building at Fifth Avenue and 38th Street in Manhattan—and I could see the entire history of twentieth-century efforts in maximal height rising before me at a single glance.

As a patriotic native and an architecture buff, I was thrilled. The world's tallest buildings, one after the other: the Flatiron Building at Fifth and 23rd, breaking the record at three hundred feet in 1903; the Metropolitan Life Tower at Madison and 24th, clocking in at seven hundred feet in 1909; the Woolworth Building on South Broadway (792 feet in 1913); a quick turn of the head to the Chrysler Building on Lexington and 42nd (1,048 feet in 1930); facing back downtown for the greatest impact of all, the massive Empire State Building just four blocks south at Fifth and 34th, occupying almost half my viewing area (1,250 feet in 1931, with extension to 1,475 feet by a TV tower installed in 1951); and finally the Twin Towers of the World Trade Center, way downtown and small by visual comparison in the distance (1,350 feet in 1976). After this, I understand that some building in Chicago went even higher, but no real New Yorker would acknowledge such a travesty.

Such an ever-rising sequence of Excelsior might give the false impression of potential extension without limit. But we should really draw the diametrically opposite conclusion that each new contender "stretches the envelope" of severe constraint; perhaps people do, but buildings, like trees, cannot reach heaven. Each higher push is an engineering marvel, straining the limits of available technology. And the increments of increase decline with time, just as improvement in sports records tails off as humans approach the right wall of biomechanical limitation (see Part Three). The Met Life Tower of 1909 more than doubled the previous record. The last few champions have added less than 10 percent to the height of the previous record holder.

In this chapter, I want to discuss the most powerful putative case for worrying about right walls in the history of human life—the saga of cultural change through time. I discussed, in the preceding chapters in Part Four, why the basic character of natural, or Darwinian, evolution—a process whose causes only produce local adaptation, not general progress—can only engender a passive trend to greater complexity in the form of a small right tail that cannot wag the dog of life's main weight at a constant bacterial mode. In this context, the issue of right walls hardly comes up—because they exist in some far uncharted distance, and life as a whole has not yet been seriously impacted by them (although individual lineages often encounter specific limits of biomechanical and other kinds of constraint—the tree that cannot reach heaven).

But human cultural change is an entirely distinct process operating under radically different principles that do allow for the strong possibility of a driven trend to what we may legitimately call "progress" (at least in a technological sense, whether or not the changes ultimately do us any good in a practical or moral way). In this sense, I deeply regret that common usage refers to the history of our artifacts and social organizations as "cultural evolution." Using the same term—evolution—for both natural and cultural history obfuscates far more than it enlightens. Of course, some aspects of the two phenomena must be similar, for all processes of genealogically constrained historical change must share some features in common. But the differences far outweigh the similarities in this case. Unfortunately, when we speak of "cultural evolution," we unwittingly imply that this process shares essential similarity with the phenomenon most widely described by the same name—natural, or Darwinian, change. The common designation of "evolution" then leads to one of the most frequent and portentous errors in our analysis of human life and history—the overly reductionist assumption that the Darwinian natural paradigm will fully encompass our social and technological history as well. I do wish that the term "cultural evolution" would drop from use. Why not speak of something more neutral and descriptive—"cultural change," for example?

The obvious main difference between Darwinian evolution and cultural change clearly lies in the enormous capacity that culture holds—and nature lacks—for explosive rapidity and cumulative directionality. In an unmeasurable blink of a geological eyelash, human cultural change has transformed the surface of our planet as no event of natural evolution could ever accomplish at Darwinian scales of myriad generations. (Natural catastrophes of a physical nature, like the bolide that triggered the great Cretaceous extinction, may wipe out many forms of life in a geological moment, but no known process can produce natural evolutionary change at anything like the speed and extent of human cultural transformation; the impressive and maximal rapidity of the Cambrian explosion lasted some 5 million years.)

The most impressive contrast between natural evolution and cultural change lies embedded in the major fact of our history. We have no evidence that the modal form of human bodies or brains has changed at all in the past 100,000 years—a standard phenomenon of stasis for successful and widespread species, and not (as popularly misconceived) an odd exception to an expectation of continuous and progressive change. The Cro-Magnon people who painted the caves at Lascaux and Altamira some fifteen thousand years ago are us—and one look at the incredible richness and beauty of this work convinces us, in the most immediate and visceral way, that Picasso held no edge in mental sophistication over these ancestors with identical brains. And yet, fifteen thousand years ago, no human social grouping had produced anything that would conform with our standard definition of civilization. No society had yet invented agriculture; none had built permanent cities. Everything that we have accomplished in the unmeasurable geological moment of the last ten thousand years—from the origin of agriculture to the Sears Building in Chicago, the entire panoply of human civilization for better or for worse—has been built upon the capacities of an unaltered brain. Clearly, cultural change can vastly outstrip the maximal rate of natural Darwinian evolution.

Among the many differences in deep principle between natural evolution and cultural change, two stand out as the motors of cultural rapidity and directionality:

1. TOPOLOGY.   Darwinian evolution at the species level and above is a story of continuous and irreversible proliferation. Once a species (defined by inability to reproduce with members of any other species) becomes separate from an ancestral line, it remains distinct forever. Species do not amalgamate or join with others. Species interact in a rich variety of ecological ways, but they cannot physically join into a single reproductive unit. Natural evolution is a process of constant separation and distinction.

Cultural change, on the other hand, receives a powerful boost from amalgamation and anastomosis of different traditions. A clever traveler may take one look at a foreign wheel, import the invention back home, and change his local culture fundamentally and forever. One brace of guns, one bevy of war chariots, imported with the engineers and tradesmen to keep them in working order, can transform a limited and peaceful state into an expanding engine of conquest. The explosively fruitful (or destructive) impact of shared traditions powers human cultural change by a mechanism unknown in the slower world of Darwinian evolution.

2. MECHANISM OF INHERITANCE.   Darwinian evolution works by the indirect and inefficient mechanism of natural selection. Effectively random variation must first provide the raw material of change, and natural selection—a negative force that can make nothing by itself—then acts by eliminating most variants and preserving those individuals fortuitously better adapted to changing local environments. The summation of favorable variants over many generations leads to evolutionary change. Local improvement rises upon the hecatomb of countless deaths; we get to a "better" place by removing the ill-adapted, not by actively constructing an improved version.

Anyone can easily envision a more direct and efficient mechanism: Why can't organisms figure out what would do them good, develop those adaptive features by dint of effort during their lifetimes, and then pass those improvements to their offspring in the form of altered heredity? We call such a putative mechanism "Lamarckism," or "the inheritance of acquired characters." Natural evolution would go like gangbusters if heredity happened to work in this manner. But, unfortunately, it doesn't. Inheritance is Mendelian, not Lamarckian. An organism may struggle to "improve" throughout its life—the giraffe stretching its neck upward, or the blacksmith developing a strong right arm, to cite the clichéd and ridiculous examples of our schoolday textbooks—but these advantageous "acquired characters" cannot be passed to offspring because they do not alter the genetic material that will build the next generation. Too bad, but so be it. Darwinism works well enough, if slowly and indirectly.

But cultural change, on a radical other hand, is potentially Lamarckian in basic mechanism. Any cultural knowledge acquired in one generation can be directly passed to the next by what we call, in a most noble word, education. If I invent the first wheel, my brainchild is not condemned to oblivion by hereditary impassability (as any purely bodily improvement would be). I just teach my children, my apprentices, my social group, how to make more wheels. The point is so simple, yet so profound. Reading, writing, filming, teaching, practicing, apprenticing, learning— all the distinctly human activities that pass knowledge between generations—act as the Lamarckian boosters of our cultural history. This uniquely and distinctively Lamarckian style of human cultural inheritance gives our technological history a directional and cumulative character that no natural Darwinian evolution can possess.

The net result of these two crucial differences between natural evolution and cultural change—the enormous boosts given to the cultural mode by amalgamation of lineages and Lamarckian inheritance—also specifies a key distinction crucially relevant to the central theme of this book. Natural evolution includes no principle of predictable progress or movement to greater complexity. But cultural change is potentially progressive or self-complexifying because Lamarckian inheritance accumulates favorable innovations by direct transmission, and amalgamation of traditions allows any culture to choose and join the most useful inventions of several separate societies.

I should introduce the obvious caveat at this point. A potential for inherent "progress" provides no guarantee of realization in actuality. The radical contingency of all history can intervene in a thousand potential ways. A capacity for technological accumulation does not guarantee that all cultures will avail themselves of this potentially mixed blessing. In fact, several great societies have made conscious decisions not to pursue technological "progress" to the inevitable destruction of an old order. At a crucial point in the history of human life, imperial China decided to scrap the technology of interoceanic shipping and navigation that, if pursued, might well have converted the central historical theme of European westward expansion to an alternative tale of Oriental eastward exploration in the New World. In the early 1640s, after a century of relative openness to Western inventions, especially to the musketry that permitted their assumption and consolidation of power, Japan's ruling Tokugawa shogunate severed all future accumulation and banned most of what had been imported. So complete and sudden was the cutoff that Japanese inhabitants of various trading cities established abroad were not even allowed to return home. All Western trade was reduced to the merest trickle. Only two Dutch ships could arrive each year. They could dock only at Nagasaki, and all Dutch traders had to live on the artificial island of Dejima, connected to the rest of Nagasaki by a narrow and easily guarded causeway.

Moreover, and obviously, accumulating technological "progress" need not lead to cultural improvement in any visceral or moral sense—and may just as well end in destruction, if not total extinction, as various plausible scenarios, from nuclear holocaust to environmental poisoning, suggest. I have long been impressed by a potential solution, perhaps whimsically proposed, but worthy of serious attention in my view, to the problem of why we haven't been contacted by the plethora of advanced civilizations that ought to inhabit other solar systems in our universe. Perhaps any society that could build a technology for such interplanetary, if not intergalactic, travel must first pass through a period of potential destruction where technological capacity outstrips social or moral restraint. And perhaps no, or very few, societies can ever emerge intact from such a crucial episode.

Nonetheless, and despite this important caveat on the difference between technological complexification and a proper vernacular sense of progress or human good, I must still reassert the bearing of the crucial difference between cultural change and natural evolution upon the central theme of this book: Cultural change operates by mechanisms that can validate a *general and driven trend to technological progress*—so very different from the minor and passive trend that Darwinian processes permit in the realm of natural evolution. And once you start to operate by general and driven trends, you can move very deliberately, and very fast. With directed motion of this sort, you ought to start running into right walls. Thus, as one crucial difference in interpreting our cultural history versus the natural evolution of life, our institutions should frequently be shaped and troubled by right walls (I have already given one example from the history of batting averages in baseball), whereas life's evolution, with its massive and persistent bacterial mode, and puny right tail, should rarely encounter this bounding theme of full houses. Let us therefore consider three important aspects of our cultural life (I invite readers to contemplate many others, here omitted for no reason beyond my own limitations) that may be impacted quite differently by relevant right walls.

1. SCIENCE. God bless ignorance! If we were much smarter, or had been at it much longer, we might actually be approaching a right wall of complete (or at least adequate) knowledge, thus leaving scientists with little of interest to do. We are in no danger whatever of any such limitation over the next several generations. In other words, our current storehouse of knowledge lies so far from the right wall of what we might learn that science need not fear any obsolescence.

I do not, of course, say that all subfields remain forever open, or that we can never reach completion for certain circumscribed aspects of natural reality—but only that any closure leaves so many adjacent open fields that no good intellect need ever fear superannuation. For example, if you have a passion for describing new species of birds, then a right wall might stymie your desires, for nearly all of the earth's eight thousand or so species have probably now been found and described. But why not switch to beetles, where you need never fear completion amid several hundred thousand already named, and probably a few million still undescribed?

I don't mean to be entirely whimsical or completely sanguine. Certain victories in the game of knowledge are so sweet, so pervasive in their impact, so defining of a profession, that we can hardly hope to equal their importance within the same world of discourse. As a graduate student, I watched plate tectonics sweep through my field of geology. These were exciting times indeed, but who could ever match the thrill of an earlier discovery, vouchsafed to geologists of the late eighteenth and early nineteenth centuries, that time comes in billions (as we now know) rather than thousands of years. Once geology grasped this great reform, no other intellectual reconstruction could ever again be so vast. And whatever the excitement and pleasure of new discoveries made every year by biologists, no one will ever again experience the ultimate intellectual high of reconstructing all nature with the passkey of evolution—a privilege accorded to Charles Darwin, and now closed to us. But we have so much to do, so much to understand, so many puzzles so far from solution, that we cannot even conceive their formulation under constraints of our present worldview. So why worry about right walls?

2. THE PERFORMING ARTS.   In this domain, above all others, our best practitioners probably stand closest to right walls of human limitation—especially for any activity involving bodily strength and dexterity that has been practiced with great potential reward (thereby attracting the best candidates for sustained excellence) during a long period of time. I suspect that our very best performers have long stood about as close to the wall as humans are likely to get for several important activities. Consider musical performance on instruments of relatively unchanged design. I doubt that Isaac Stern plays better than Paganini, Vladimir Horowitz than Liszt, or E. Power Biggs than Bach. In some respects, particularly for lost skills and changing sensibilities, we may now be worse off. Can anyone today sing like Farinelli? Can anyone (or, in this case, could anyone) ever play the natural horn (precursor to the difficult, but playable, French horn) without embarrassing errors?

In sports, as discussed in Part Three, some records continue to decline substantially, especially for activities (like women's track and field) only recently encouraged and honored. But the near stability, or only very slow improvement, in other records indicates a present position close to a limiting right wall.

And yet, although we stand so close to a wall for many activities in the performing arts, I don't think that we find the implied limits troubling—for two reasons based on perceptions about the nature of this enterprise held both by performers and spectators.

First, we don't demand transcendence in the performing arts. Repetition of maximal excellence is entirely permissible. We don't expect Pavarotti to sing better every time; and we don't expect Tony Gwynn to raise his batting average every season. When we thrill to Isaac Stern's rendition of Beethoven's violin concerto, we are not bothered by the probability that Paganini played the same piece just as well more than a century ago. Our standards, in other words, are absolute, not relative. Since so few people can ever get there, we honor any performance, at any time, that touches the divine realm of the right wall of human limits. A performer just has to exist in this region; he needn't improve upon his past perfection, or exceed someone else's surpassingly rare achievement.

Second, humans have a remarkable capacity to scale their expectations and excitements to the character of the enterprise. When maximal performance stands about a football field from a wall, only improvements measured in yards will be viewed as impressive. But when the best reside only one millimeter from a wall, even a measurable micron of improvement will send devotees into swoons of rapture.

This drive to betterment, this internal need to shave a micron, probably affects performers more than spectators—for many improvements in this category are entirely invisible to all but the most discerning spectator, while performers will often literally die for a chance at minuscule transcendence. If this isn't divine madness, then such a sublime concept has no meaning. So long as the best of us are driven to seek heights of excellence, to stretch the proverbial envelope no matter how little, to regard compromise as beyond contemplation, there is hope for humanity.

My favorite examples come from a discipline that, perhaps more than any other, drives the best practitioners to a never-ending search for transcendence in realms already operating close to physical and biomechanical limits of Newtonian existence—circus performers. Only so many balls can be juggled aloft; a body can make only so many turns in the air before the speed of plunge defies any attempt by a catcher to hold his trapeze partner in descent.

Jules Léotard invented the flying return trapeze in 1859. No one managed to perform the supposedly impossible triple somersault to a catcher until 1897, though several performers died in the attempt (the driven and the foolhardy often refused to perform with nets—and one can also break a neck by falling badly into a net). Only in the 1930s did a trapeze artist, the great Alfredo Codona, manage to perfect the triple as a standard act (he succeeded about nine times out of ten, as his body flew through the air at sixty miles per hour in the height of his plunge to a catcher). Codona wrote of his quest:

The history of the triple somersault is a history of death;
as long as there have been circuses, there have been men
and women whose sole ambition was to accomplish three
full turns in the air. The struggle to master it has lasted
more than a century, beginning with the old days of the
famous leapers who worked with a springboard, and the
triple somersault has killed more persons than all other
dangerous circus acts combined.

Subsequent history illustrates the joy and frustration of pushing envelopes toward a nearby right wall of strict limits. In 1982, Miguel
Vasquez, flying at seventy-five miles an hour to a catcher, his brother
Juan, first threw a quadruple somersault in public performance. Only a
few aerialists have succeeded since then, and no one has been able to perform the quadruple consistently (I have witnessed five attempts—all failures, and several by the Vasquez brothers). But the passion for
transcendence continues. On December 30, 1990, *The New York Times
Magazine* featured a long article on the quest, not yet successful, by a Russian group to perform the quintuple.

The number of people who can balance in set configurations on a high
wire should be defined by laws of physics, but great performers continue
to pursue the impossible (and end up in glory at the right wall, or in death).
Karl Wallenda, the greatest wire walker in history, drilled his whole family in the art and constantly sought new achievements deemed impossible. One admirer wrote (Hammarstrom, 1980, page 48): "Some people
thought the great Wallenda was crazy; I think he was incredible." Wallenda perfected the seven-person pyramid on the high wire, but one night
in Detroit the configuration collapsed as the lead man fell. Two performers died and a third was paralyzed. Wallenda himself, age seventy-three,
died in Puerto Rico on March 22, 1978, when a strong gust of wind blew
him off a wire strung from the tenth story between two beachfront hotels.

3. THE CREATIVE ARTS. If science stands too far from a right wall to
worry about limitation, and if great performers nearly touch the wall but
do not feel diminished by restricted domains of potential improvement,
then a third category of creative arts does face a potentially painful

dilemma based on our decision to adopt an ethic of innovation that awards greatness only to those who devise a novel style (a criterion not always followed in Western history, but very strong at the moment).

Suppose that the mile run had disappeared as a competitive sport as soon as a hundred people covered the distance in less than four minutes. Given an ethic that exalts perennial originality in style of artistic composition, the history of classical music (and several other arts) may fall into such a domain. One composer may exploit a basic style for much of a career, but successors may not follow this mode in much detail, or for very long. This perpetual striving for novelty may grant us joy forever if a limitless array of potential styles awaits discovery and exploitation. But perhaps the world is not so bounteous. Perhaps we have already explored most of what even a highly sophisticated audience can deem accessible. Perhaps, in other words, we have reached the right wall of styles that a sympathetic, intelligent, but still nonprofessional audience can hope to grasp with understanding and compassion.

The standard retort of artists to charges of inaccessibility has become such a mantra that any questioner gets quickly dismissed as a hopeless Philistine: "The complaint could only be made by a pitiful, dried-up old guard. They said the same thing of Beethoven, and of Van Gogh. The future will vindicate us. The cacophony of today will be hailed as a grand innovation tomorrow." As Beethoven said to a conservative musician who wondered out loud whether his Razoumovsky Quartets could be defined as music: "They aren't for you, but for a later age."

Fine. Sometimes. But will this claim always wash, and should we regard its venerable status as above criticism? I think that we should take the argument of the right wall as a serious alternative: perhaps the range of accessible styles can become exhausted, given the workings of human neurology and the consequent limits of understanding. Perhaps we can reach a right wall of potential popularity, where our continued adherence to an ethic of innovation effectively debars newcomers, whatever their potential talents, from becoming the Mozart of the new millennium.

I don't know how else to resolve what I like to call the "German virus problem." Between 1685 (the birth of Bach and Handel) and 1828 (the death of Schubert), the small world of German-speaking people gave us the full life spans of Bach, Handel, Haydn, Mozart, Beethoven, and Schubert, to mention just a few. Where are their counterparts today? Who, in the vastly larger domain of the entire world, with musical training available to so many million more people, would you choose among late-twentieth-century composers to rank with these men?

I can't believe that a musical virus, now extinct, was then loose in the German-speaking world. Nor can we deny that many more people of equal or greater potential talent must now be alive and active somewhere on this planet. What are they doing? Are they writing in styles so arcane that only a rarefied avant-garde of professionals has any access? Are they performing jazz, or (God help us) rock, or some other genre instead? I do suspect that these people exist, but are victims of the right wall and our unforgiving ethic of innovation.

I don't have any solutions to propose. I don't think that we should find these folks and let them master an old style to write Beethoven's Tenth Symphony or compose Mozart's opera on the tragedy of King Lear. I do understand why such an activity might be deemed unappealing. Nonetheless, I do think that we should face the problem and rethink some knee-jerk notions about novelty above all, and future accessibility for anything.

Finally, what major lesson can we learn from the general model of Full House—the focus on variation as an ultimate reality, and the relegation of means and extremes to a realm of Platonic abstraction (sometimes useful, but always less than the whole)? I like to think of myself as a tough-minded intellectual, a foe of all fuzziness from alien abductions to past-life regressions. I hate to think that an intellectual position, hopefully well worked out in the pages of this book, might end up as a shill for one of the great fuzzinesses of our age—so-called "political correctness" as a doctrine that celebrates all indigenous practice, and therefore permits no distinctions, judgments, or analyses.

And yet I think that the Full House model does teach us to treasure variety for its own sake—for tough reasons of evolutionary theory and nature's ontology, and not from a lamentable failure of thought that accepts all beliefs on the absurd rationale that disagreement must imply disrespect. Excellence is a range of differences, not a spot. Each location on the range can be occupied by an excellent or an inadequate representative—and we must struggle for excellence at each of these varied locations. In a society driven, often unconsciously, to impose a uniform mediocrity upon a former richness of excellences—where McDonald's drives out the local diner, and the mega–Stop & Shop eliminates the corner Mom and Pop—an understanding and defense of full ranges as natural reality might help to stem the tide and preserve the rich raw material of any evolving system: variation itself.

We turn, with fascination and respect, to the lines that Darwin carefully chose to end his revolutionary book, the *Origin of Species*. He did not celebrate evolution by lauding the development of human intelligence or any upward march to preordained and preferable complexity. Rather, he chose to honor life's bursting and bustling variety in contrast with the dull repetition of earthly revolution about the sun in all its Newtonian majesty (he also acknowledged life's beginning at the left wall):

> Whilst this planet has gone cycling on according to the fixed law of gravity, from so simple a beginning endless forms most beautiful and most wonderful have been, and are being, evolved.

He began these final lines with the best epitome of all: "There is grandeur in this view of life."

Bibliography

• • •

Adams, D. 1981. The probability of the league leader batting .400. *Baseball Research Journal*, 82–83.

Arnold, A. J., D. C. Kelly, and W. C. Parker. 1995. Causality and Cope's Rule: evidence from the planktonic Foraminifera. *Journal of Paleontology*, 69:203–10.

Augusta, J., and Z. Burian, 1956. *Prehistoric Animals.* London: Spring Books.

Baross, J. A., M. D. Lilley, and L. I. Gordon. 1982. Is the CH4, H2, and CO venting from submarine hydrothermal systems produced by thermophilic bacteria? *Nature,* 298:366–68.

Baross, J. A., and J. W. Deming. 1983. Growth of "black smoker" bacteria at temperatures of at least 250°C. *Nature,* 303:423–26.

Boyajian, G., and T. Lutz. 1992. Evolution of biological complexity and its relation to taxonomic longevity in the Ammonoidea. *Geology,* 20:983–86.

Broad, W. J. 1993. Strange new microbes hint at a vast subterranean world. *The New York Times,* 28 December, C1.

Broad, W. J. 1994. Drillers find lost world of ancient microbes. *The New York Times,* 4 October, C1.

Brown, J. H., and B. A. Maurer. 1986. Body size, ecological dominance, and Cope's rule. *Nature*, 324:248–50.

Carew, R., and I. Berkow. 1979. Carew. New York: Simon and Schuster.

Chatterjee, S., and M. Yilmaz. 1991. Parity in baseball: stability of evolving systems. Draft manuscript.

Chen J.-Y., J. H. Dzik, G. D. Edgecombe, L. Ramsköld, and G.-Q. Zhou. 1995. A possible early Cambrian chordate. *Nature* 377:720–22.

Cope, E. D. 1896. *The primary factors of organic Evolution.* Chicago: The Open Court Publishing Company.

Curran, W. 1990. *Big Sticks: The Batting Revolution of the Twenties.* New York: William Morrow and Company.

Dana, J. D. 1876. *Manual of Geology,* Second Edition. New York: Ivison, Blakeman, Taylor and Company.

Darwin, C. R. 1859. *On the Origin of Species.* London: John Murray.

Durslag, M. 1975. Why the .400 hitter is extinct. *Baseball Digest,* August, 34–37.

Eckhardt, R. B., D. A. Eckhardt, and J. T. Eckhardt. 1988. Are racehorses becoming faster? *Nature,* 335:773.

Eldredge, N., and S. J. Gould. 1972. Punctuated equilibria: An alternative to phyletic gradualism. In T. J. M. Schopf, ed., *Models in Paleobiology,* 82–115. San Francisco: Freeman, Cooper & Company.

Fellows, J., P. Palmer, and S. Mann. 1989. On the tendency toward increasing specialization following the inception of a complex system—professional baseball 1871–1988. Draft manuscript.

Figuier, L. 1867. *The World Before the Deluge*: A New Edition. London: Chapman & Hall.

Fuhrman, J. A., K. McCallum, and A. A. Davis. 1992. Novel major archaebacterial group from marine plankton. *Nature,* 356:148–49.

Fuhrman, J. A., T. D. Sleeter, C. A. Carlson, and L. M. Proctor. 1989. Dominance of bacterial biomass in the Sargasso Sea and its ecological implications. *Marine Ecology Progress Series,* 57:207–17.

Gilovich, T., R. Vallone, and A. Tversky. 1985. The hot hand in basketball: On the misperception of random sequences. *Cognitive Psychology,* 17:295–314.

Gingerich, P. D. 1981. Variation, sexual dimorphism, and social structure in the early Eocene horse Hyracotherium (Mammalia, Perissodactyla). *Paleobiology* 7:443–55.

Gold, T. 1992. The deep, hot biosphere. *Proceedings of the National Academy of Sciences* USA, 89:6045–49.

Gould, S. J. 1983. Losing the edge: the extinction of the .400 hitter. *Vanity Fair*, March, 120, 264–78.

Gould, S. J. 1985. The median isn't the message. *Discover,* June, 40–42.

Gould, S. J. 1986. Entropic homogeneity isn't why no one hits .400 any more. *Discover,* August, 60–66.

Gould, S. J. 1987. Life's little joke; the evolutionary histories of horses and humans share a dubious distinction. *Natural History,* April, 16–25.

Gould, S. J. 1988. The case of the creeping fox terrier clone. *Natural History,* January, 16–24.

Gould, S. J. 1988. Trends as changes in variance: a new slant on progress and directionality in evolution. *Journal of Paleontology,* 62(3):319–29.

Gould, S. J. 1988. The Streak of Streaks. *The New York Review of Books,* 35:8–12, 18 August.

Gould, S. J. 1989. *Wonderful Life: The Burgess Shale and the Nature of History.* New York: W.W. Norton.

Gould, S. J. 1991. The birth of the two-sex world. Review of "Making sex: body and gender from the Greeks to Freud," by Thomas Laqueur. *The New York Review of Books,* 38:11–13, 13 June.

Gould, S. J. 1993. Prophet for the Earth. Review of "The Diversity of Life" by E. O. Wilson. *Nature,* 361:311–12.

Gould, S. J. 1995. Of it, not above it. *Nature,* 377:681–82.

Gould, S. J. 1996. Triumph of the root-heads. *Natural History,* January, 10–17.

Gould, S. J. 1996. *Dinosaur in a Haystack.* New York: Harmony Books.

Gould, S. J., and N. Eldredge. 1993. Punctuated equilibrium comes of age. *Nature,* 366: 223–27.

Gould, S. J., and R. C. Lewontin. 1979. The spandrels of San Marco and the Panglossian paradigm: A critique of the adaptationist programme. *Proceedings of the Royal Society of London Series B.* 205:581–98.

Gould, S. J., and E. S. Vrba. 1982. Exaptation—a missing term in the science of form. *Paleobiology* 8(1):4–15.

Hammarstrom, D. L. 1980. *Behind the Big Top.* New York: A. S. Barnes and Company.

Hoffer, R. 1993. Strokes of luck. *Sports Illustrated,* 28 June, 22–25.

Holmes, T. 1956. We'll never have another .400 hitter. *Sport,* February, 37–39, 87.

Huxley, T. H. 1880. On the application of the laws of evolution to the arrangement of the Vertebrata, and more particularly of the Mammalia. *Proceedings of the Zoological Society of London*, 43, 649–61.

Huxley, T. H. 1880. *The Crayfish, An Introduction to the Study of Zoology.* London: C. Kegan Paul and Company.

Jablonski, D. 1987. How pervasive is Cope's rule? A test using Late Cretaceous mollusks. *Geological Society of America, Abstracts with Programs,* 19:7, 713–14.

James, B. 1986. *The Bill James Historical Baseball Abstract.* New York: Villard Books.

Kaiser, J. 1995. Can deep bacteria live on nothing but rocks and water? *Science,* 270:377.

Knight, C. R. 1942. Parade of life through the ages. *National Geographic,* 81:2 (February), 141–84.

Laqueur, T. 1990. *Making Sex.* Cambridge, Mass.: Harvard University Press.

L'Haridon, S., A.-L. Reysenbach, P. Glénat, D. Prieur, and C. Jeanthon. 1995. Hot subterranean biosphere in a continental oil reservoir. *Nature,* 377:223–24.

MacFadden, B. J. 1984. Systematics and phylogeny of *Hipparion, Neohipparion, Nannippus,* and *Cormohipparion* (Mammalia, Equidae) from the Miocene and Pliocene of the New World. *Bulletin of the American Museum of Natural History* 179:1–196.

MacFadden, B. J. 1986. Fossil horses from "Eohippus" *(Hyracotherium)* to *Equus:* Scaling, Cope's law, and the evolution of body size. *Paleobiology,* 12:4, 355–69.

MacFadden, B. J., and R. Hulbert, Jr. 1988. Explosive speciation at the base of the adaptive radiation of Miocene grazing horses. *Nature,* 336:6198, 466–68.

MacFadden, B. J., and J. S. Waldrop. 1980. *Nannippus phlegon* (Mammalia, Equidae) from the Pliocene (Blancan) of Florida. *Bulletin of the Florida State Museum Biological Sciences,* 25:1, 1–37.

Margulis, L., and D. Sagan. 1986. *Microcosmos.* New York: Simon and Schuster.

Matthew, W. D. 1903. *The Evolution of the Horse.* American Museum of Natural History pamphlet.

Matthew, W. D. 1926. The evolution of the horse: A record and its interpretation. *Quarterly Review of Biology,* 1(2):139–85.

Mayr, E. 1963. *Animal Species and Evolution.* Cambridge, Mass.: Belknap Press of Harvard University Press.

McShea, D. W. 1992. A metric for the study of evolutionary trends in the complexity of serial structures. *Biological Journal of the Linnean Society of London,* 45:39–55.

McShea, D. W. 1993. Evolutionary change in the morphological complexity of the mammalian vertebral column. *Evolution,* 47:730–40.

McShea, D. W. 1994. Mechanisms of large-scale evolutionary trends. *Evolution,* 48:1747–63.

McShea, D. W. 1996. Metazoan complexity and evolution: is there a trend? *Evolution,* in press.

McShea, D. W., B. Hallgrimsson, and P. D. Gingerich. 1995. Testing for evolutionary trends in non-hierarchical developmental complexity. Abstracts, *Annual Meeting of the Geological Society of America,* New Orleans, A53–A54.

Nealson, K. H. 1991. Luminescent bacteria as symbiotic with entomopathogenic nematodes. In L. Margulis and R. Fester, eds., *Symbiosis as a Source of Evolutionary Innovation,* 205–18. Cambridge, Mass.: MIT Press.

Oliwenstein, L. 1993. Onward and upward? *Discover,* June, 22–23.

Parkes, J., and J. Maxwell. 1993. Some like it hot (and oily). *Nature,* 365:694–95.

Parkes, R. J., B. A. Cragg, S. J. Bale, J. M. Getliff, K. Goodman, P. A. Rochelle, J. C. Fry, A. J. Weightman, and S. M. Harvey. 1994.

Deep bacterial biosphere in Pacific Ocean sediments. *Nature*, 371:410–13.

Peck, M. Scott. 1978. *The Road Less Traveled.* New York: Simon & Schuster.

Prothero, D. R., E. Manning, and C. B. Hanson. 1986. The phylogeny of the Rhinocerotoidea (Mammalia, Perissodactyla). *Zoological Journal of the Linnean Society,* 87:341–66.

Prothero, D. R., and N. Shubin. 1989. The evolution of Oligocene horses. In: D. R. Prothero and R. M. Schoch, eds., *The Evolution of Perissodactyls,* 142–75. Oxford: Oxford University Press.

Prothero, D. R., C. Guérin, and E. Manning. 1989. The history of the Rhinocerotoidea. In D. R. Prothero and R. M. Schoch, eds., *The Evolution of Perissodactyls,* 321–40. New York: Oxford University Press.

Prothero, D. R., and R. M. Schoch. 1989. Origin and evolution of the Perissodactyla: summary and synthesis. In D. R. Prothero and R. M. Schoch, eds., *The Evolution of Perissodactyls,* 504–37. New York: Oxford University Press.

Richards, R. J. 1992. *The Meaning of Evolution.* Chicago: University of Chicago Press.

Rudwick, M. J. S. 1992. *Scenes from Deep Time.* Chicago: University of Chicago Press.

Sagan, D., and L. Margulis. 1988. *Garden of Microbial Delights.* New York: Harcourt Brace Jovanovich.

Simpson, G. G. 1951. *Horses.* Oxford: Oxford University Press.

Sober, E. 1984. *The Nature of Selection.* Cambridge, Mass.: MIT Press.

Stanley, S. M. 1973. An explanation for Cope's rule. *Evolution,* 27:1–26.

Stauffer, R. C. (ed.). 1975. *Charles Darwin's Natural Selection.* Cambridge, UK: Cambridge University Press.

Stetter, K. O., R. Huber, E. Blöchl, M. Kurr, R. D. Eden, M. Fielder, H. Cash, and I. Vance. 1993. Hyperthermophilic Archaea are thriving in deep North Sea and Alaskan oil reservoirs. *Nature,* 365:743–45.

Stevens, T. O., and J. P. McKinley. 1995. Lithautotrophic microbial ecosystems in deep basalt aquifers. *Science,* 270:450–54.

Szewzyk, R., M. Szewzyk, and T.-A. Stenström. 1994. Thermophilic, anaerobic bacteria isolated from a deep borehole in granite in Sweden. *Proceedings of the National Academy of Sciences USA,* 91:1810–13.

Tax, Sol (ed.). 1960. *Evolution After Darwin,* 3 volumes. Chicago: University of Chicago Press.

Thomas, R. D. K. 1993. Order and disorder in the evolution of biological complexity. Draft manuscript.

Vetter, R. D. 1991. Symbiosis and the evolution of novel trophic strategies: thiotrophic organisms at hydrothermal vents. In L. Margulis and R. Fester, eds., *Symbiosis as a Source of Evolutionary Innovation.* Cambridge, Mass.: MIT Press, 219–45.

Vrba, E. S. 1980. Evolution, species and fossils: how does life evolve? *South African Journal of Science,* 76:61–84.

Vrba, E. S., and N. Eldredge. 1984. Individuals, hierarchies and processes: towards a more complete evolutionary theory. *Paleobiology,* 10:146–71.

Walsby, A. E. 1983. Bacteria that grow at 250°C. *Nature,* 303:381.

Whipp, B. J., and S. A. Ward. 1992. Will women soon outrun men? *Nature,* 335:25.

Williams, G. C. 1966. *Adaptation and Natural Selection.* Princeton, N.J.: Princeton University Press.

Williams, T., and J. Underwood. 1986. *The Science of Hitting.* New York: Simon and Schuster.

Wilson, E. O. 1992. *The Diversity of Life.* Cambridge, Mass.: Harvard University Press.

Woese, C. R. 1994. Universal phylogenetic tree in rooted form. *Microbiological Reviews,* 58:1–9.

Yoon, C. K. 1993. Biologists deny life gets more complex. *The New York Times,* 30 March, C1.

Index
• • •

Page numbers in italic indicate illustrations. Page numbers followed by *n* refer to footnote on indicated page.

Index

Index